Beyond the Interface

How the Intersections of New Science Are Starting to Explain the Unexplainable

By

Kevin L. Meyer

Beyond the Interface: How the Intersections of New Science Are Starting to Explain the Unexplainable

Copyright © 2025 by Kevin L. Meyer

Bhavana Press
Morro Bay, California

ISBN: 979-8-9997611-6-3

First Edition

Printed in the United States of America

Contents

Part III - Toward a New Inquiry

Introduction: Through the Interface

On a cold night in 2001, cardiologist Pim van Lommel revived a patient who had been in cardiac arrest—his heart stopped, blood pressure undetectable—for more than two minutes. When the man regained consciousness, he calmly described details of the operating room he could not have seen while unconscious, including the specific brand of defibrillator pads and conversations among the medical staff.[1] The case became part of van Lommel's landmark study in *The Lancet*, the first major prospective research on near-death experiences in cardiac arrest survivors.

Months later, quantum physicists published experimental results showing that photons seemed to "choose" their behavior based on measurements that would be made in the future—a phenomenon known as quantum delayed-choice experiments, first theorized by John Archibald Wheeler and later confirmed by teams including Alain Aspect.[2] The particles appeared to retroactively determine their own past states.

And in laboratories studying collective behavior, researchers recorded starlings in flight demonstrating coordination so precise and instantaneous it challenged conventional models of information transfer. The birds moved as a unified entity, with behavioral changes propagating across flocks faster than any known signaling mechanism could explain.[3]

Different stories. Different disciplines. Yet each hints at the same fundamental puzzle: the world we experience through our senses may not capture the full nature of reality.

The Nature of the Interface

Every moment of human experience unfolds at a dynamic boundary—what we might call *the interface*—where mind meets world. We are not passive recipients of an objective reality but active constructors of experience. Our brains constantly generate predictions about incoming sensory data, updating these models based on error signals when expectations don't match input.[^4] This process, known as predictive coding, means that much of what we "perceive" is actually our brain's best guess about what's happening around us.

Like a computer's graphical interface, our perceptual system hides the underlying complexity. We never directly experience photons striking retinal cells or electrical impulses traveling along neural pathways—we see a sunset. We do not register compression waves in air molecules—we hear music. The interface translates the incomprehensible complexity of physical reality into the coherent, meaningful world of human experience.

This translation process is now understood to be far more active and creative than previously imagined. Leading neuroscientists like Karl Friston describe the brain as a "prediction machine" that constantly generates hypotheses about reality and updates them based on incoming evidence.[^5] Lisa Feldman Barrett's research suggests that even emotions are constructed through this predictive process rather than being hardwired responses.[^6]

Understanding the interface as an active, constructive process raises profound questions: What aspects of reality might our evolved perceptual systems systematically miss or distort? How might phenomena that seem impossible within our everyday experience actually reflect limitations of the interface itself?

Why This Book Exists

I am not a laboratory researcher tied to a single discipline. My work is that of a synthesizer—connecting insights across fields that rarely communicate, finding what I call the "Venn zone" where different domains of knowledge overlap to reveal richer understanding. This approach, which I term **Venn Thinking**, maps intersections between neuroscience, physics, psychology, anthropology, and other fields to explore questions that no single discipline can address alone.

There is a significant gap in how we approach mysterious phenomena. Popular books on consciousness and anomalous experiences often lack scientific rigor, relying on cherry-picked anecdotes or speculative leaps. Academic works, conversely, tend to avoid edge cases entirely—dismissing them as too anecdotal, too culturally complex, or too risky for professional reputations. Mainstream scientists and dedicated skeptics both play essential roles in advancing knowledge, but the conversation between them often stagnates in predictable positions.

This book occupies the middle ground: maintaining rigorous standards while remaining open to phenomena that challenge current paradigms. My methodology follows strict criteria:

- **Primary Sources**: Peer-reviewed studies, established researchers, and documented historical records form the foundation for every claim
- **Cultural Context**: Every phenomenon is examined within its historical and cross-cultural setting
- **Balanced Perspective**: Both proponents and critics receive fair hearing, with particular attention to published reviews and meta-analyses that reflect genuine scientific consensus
- **Evidence Threshold**: To merit inclusion, a phenomenon must show consistent patterns across cultures, ongoing scientific investigation, and measurable characteristics that can be studied

This is not "social media science" or wishful thinking disguised as inquiry. Every assertion in these pages traces to credible, verifiable sources.

The goal is not to settle debates but to model productive investigation—equipping readers with both analytical tools and the intellectual humility necessary for exploring the boundaries of current knowledge.

Why This Book Now

We live at a unique moment when disciplinary boundaries are becoming increasingly porous. Advances in neuroimaging allow us to observe the predictive brain in action. Quantum information theory informs not only physics but cognitive science. Artificial intelligence research reveals surprising insights about consciousness and perception. Complex systems theory explains emergent behaviors from starling flocks to neural networks.

This convergence creates new opportunities for understanding phenomena that previously seemed isolated in their respective fields. In *Part I*, we will explore how the interface operates, drawing on recent discoveries about predictive processing, quantum foundations, consciousness theories, and artificial intelligence. These chapters establish scientific frameworks for approaching mystery with appropriate rigor.

In *Part II*, we venture into more contested territory: near-death experiences during cardiac arrest, children's apparent memories of previous lives, precognitive dreams, telepathic communication, and encounters with entities during psychedelic states. Much of this material exists either in technical journals or has been diluted into popular simplifications. This book aims to bridge that gap—neither dismissing phenomena that deserve investigation nor abandoning critical thinking in favor of uncritical wonder.

Current mysteries like dark matter in physics or the hard problem of consciousness in neuroscience remind us that the scientific revolution continues. What seems impossible today may become commonplace tomorrow, just as quantum mechanics and relativity once challenged intuitions about reality.

A Reader's Contract

This book asks for neither belief nor skepticism, but sustained curiosity. It will invite you to consider possibilities without rushing to judgment, to examine both strengths and limitations of competing explanations, and to recognize the current boundaries of human understanding.

We will explore domains that remain scientifically controversial. These investigations offer not proofs but data points— documented human experiences that challenge existing models. We will also examine well-established phenomena like the brain's predictive nature and complex systems behavior, considering how they might illuminate more speculative questions.

Above all, *Beyond the Interface* is about expanding conversation beyond false choices between reductive materialism and uncritical mysticism. If reality extends beyond what our evolved perceptual systems reveal, then our best approach may be learning to see the interface itself—understanding both its remarkable capabilities and its inherent limitations.

The interface that mediates our experience of reality is neither perfect nor complete. Recognizing this opens space for wonder while maintaining the intellectual rigor necessary for genuine discovery.

Prologue: What Was Once Unexplainable

Not so long ago, the world was a far stranger place. Lightning was the weapon of gods, disease the punishment for sin, and dreams messages carried across the veil by wandering spirits. Gravity was a mystical "tendency of things to fall," and the very idea of traveling faster than a horse could gallop seemed like fantasy.

Today, we can measure the voltage of a lightning strike, identify viruses under an electron microscope, and use satellites to map the curvature of spacetime. We can sequence DNA to read the code of life and cross an ocean in less time than it once took to cross a single kingdom. What was once mystery is now method.

The Pattern of the Impossible

History reveals a consistent rhythm where what seems impossible becomes commonplace through paradigm shifts in thinking rather than mere technological progress. Each transformation follows a familiar arc: observation → speculation → skepticism → explanation. The phenomena themselves don't change; our frameworks for understanding them do.

Lightning — Once divine anger wielded by Zeus or Thor, now understood as electrical discharge between regions of differing potential. Benjamin Franklin's kite experiment in 1752 transformed supernatural terror into Benjamin Franklin's practical invention of the lightning rod.[1]

Disease — For millennia attributed to curses, "bad air" (miasma), or divine punishment until Louis Pasteur's germ theory and Robert Koch's postulates in the 1870s replaced superstition with microbiology. The shift from supernatural to bacterial explanations revolutionized medicine and doubled human life expectancy.[2]

Gravity — Aristotle's explanation of falling objects as seeking their "natural place" dominated Western thought for nearly two millennia. Newton's *Principia Mathematica* (1687) reframed gravity as a universal force acting across space. Einstein's field equations (1915) later revealed it as the curvature of spacetime itself—each revolution overturning the previous "common sense."[3]

Flight — Lord Kelvin, president of the Royal Society, declared in 1895 that "heavier-than-air flying machines are impossible." Eight years later, the Wright brothers achieved powered flight at Kitty Hawk. Within fifty years, humans had broken the sound barrier and reached the moon.[4]

Space Travel — Once confined to mythology and Jules Verne's imagination, rocket technology developed by Robert Goddard and others made orbital flight routine enough to broadcast live from the International Space Station.

Each transformation required not just new observations but new ways of thinking—conceptual revolutions that made the previously unthinkable seem inevitable.

The Role of Culture in Framing the Unknown

What a culture believes about mystery shapes how phenomena are investigated—or ignored entirely. For the Norse, thunder was Thor's chariot wheels rolling across the heavens; for ancient Chinese philosophers, it was the roar of a celestial drum. Neither description was "wrong" within its cultural framework; both organized experience into meaningful patterns. But neither could lead to Franklin's lightning rod without a fundamental shift in worldview.

The same dynamic operates today. Our current explanations—quantum fields, neural correlates, spacetime curvature—are no less culturally situated than ancient mythologies. They simply represent our present interface for interpreting the unknown. Future generations may regard our certainties about consciousness, causation, and the nature of reality with the same bemused affection we reserve for medieval theories of celestial spheres.

Productive skepticism plays a crucial role in this process. Lord Kelvin's doubts about aviation, while ultimately wrong, spurred more rigorous calculations and better engineering. Similarly, contemporary skepticism about phenomena like consciousness, quantum foundations, or anomalous cognition drives deeper investigation and more precise experimental design.

What We Remember and What We Forget

Our historical record of mysteries is inevitably selective. Some phenomena receive meticulous documentation in chronicles, folklore, or sacred texts; others are ignored, dismissed, or actively erased. Ancient Egyptian physicians recorded detailed cases of head trauma but rarely documented spontaneous healing. Medieval monastic scribes preserved astronomical tables but often omitted reports of unusual aerial phenomena. Scientific journals today publish studies of neuroplasticity but rarely investigate reports of extraordinary mental capabilities.

This *reporting bias* means our understanding of historical anomalies is filtered through social acceptability, institutional priorities, and cultural blind spots. What survives in the historical record reflects not just what happened, but what various societies deemed worthy of preservation. The same interface that shapes our current perception of reality has always influenced which mysteries get investigated and which get forgotten.

Modern Mysteries: The Pattern Continues

Today's scientific frontiers remind us that the transformation from mystery to understanding remains ongoing. Dark matter and dark energy comprise 95% of the universe yet remain largely incomprehensible. The "hard problem" of consciousness—how subjective experience arises from neural activity—continues to puzzle neuroscientists and philosophers. Quantum mechanics produces reliable technology while challenging fundamental assumptions about locality, causality, and measurement.

The digital revolution has created new categories of mystery. How do large language models generate coherent responses without programmed understanding? What explains the emergence of collective intelligence in networked systems? These questions may seem as baffling to us as electricity seemed to pre-industrial societies.

An Invitation

If history teaches anything, it is that the boundary between mystery and understanding is neither fixed nor final. Many phenomena we consider routine—antibiotics eliminating deadly infections, GPS satellites coordinating global navigation, brain imaging revealing thoughts in real-time—would have been indistinguishable from magic to observers just generations ago.

The same transformation may await phenomena currently at the edges of respectability: near-death experiences during cardiac arrest, childhood memories of previous lives, telepathic communication between emotionally bonded individuals, precognitive dreams, or encounters with apparently autonomous entities during altered states of consciousness. Today's anomalies may become tomorrow's textbook science, just as yesterday's impossibilities became today's technologies.

The challenge is cultivating intellectual humility—neither dismissing unexplained phenomena nor abandoning critical thinking in favor of wishful speculation. The most productive approach combines rigorous methodology with genuine curiosity, following evidence wherever it leads while maintaining appropriate skepticism about extraordinary claims.

In the chapters ahead, we will apply this perspective to contemporary mysteries—not to prove or disprove any particular claim, but to explore what these experiences reveal about the interface between mind and world. The history of science suggests a simple truth: the impossible is often just the unexamined.

Part I

Tools at the Edge of Knowing

Real science meets the mysterious. Before exploring phenomena that challenge conventional explanation, we need the right conceptual tools. This section examines cutting-edge theories about how the brain predicts reality, how consciousness might emerge from complexity, how quantum mechanics reveals the strange nature of fundamental reality, how artificial intelligence mirrors human cognition, how mystical experiences arise across cultures, and how networks and systems generate emergent patterns. These frameworks don't provide easy answers, but they offer sophisticated ways to approach the unknown—revealing that reality itself may be far stranger than our everyday experience suggests.

Chapter 1: The Predictive Brain

Perception, Memory, and Free Will as Outputs of Generative Modeling

In 2013, researchers at the University of Glasgow presented volunteers with a deceptively simple task.[1] Participants viewed black-and-white photographs while lying inside an fMRI scanner—faces, landscapes, everyday objects. Many images depicted things the volunteers had only encountered in color: green grass, yellow bananas, red London buses. Despite the complete absence of color information on the screen, their brains responded as if color were present. Visual cortex areas specialized for color processing activated just as strongly as they would when viewing the actual colored versions.

The volunteers weren't simply remembering color—they were, in a neurologically meaningful sense, experiencing it. Their brains had generated color perception from expectation alone, filling in details that weren't there based on prior knowledge. This wasn't an unusual glitch but a window into how perception normally operates: as an active construction process rather than passive recording.

Similar experiments have revealed the brain's predictive nature across domains. When people listen to sentences with missing words, auditory cortex fires as if hearing the absent syllables.[2] When experienced drivers navigate familiar routes, their visual systems anticipate upcoming turns before the road curves into view.[3] These findings converge on a startling conclusion: much of what we experience as direct perception is actually sophisticated simulation.

This discovery has profound implications. If the brain generates so much of our experienced reality internally, what does this mean for our understanding of consciousness, free will, and the nature of subjective experience? How can we trust our perceptions if they're largely constructed rather than received?

Core Concepts: How the Brain Predicts Reality

The **predictive processing** framework, also known as predictive coding, suggests that the brain's primary function is not to react to sensory input but to anticipate it. Rather than waiting for information to arrive from the senses, the brain continuously generates predictions about what it expects to encounter. These predictions, based on prior experience and current context, form our moment-to-moment conscious experience.

Karl Friston, who has provided the most mathematically sophisticated account of this process, describes it through the **free-energy principle.**[4] According to this framework, biological systems minimize "surprise"—technically defined as the difference between expected and actual sensory input. When predictions match incoming data, processing is efficient and we experience smooth, coherent perception. When mismatches occur, the brain generates **prediction errors** that drive learning and model updating.

This process operates hierarchically throughout the nervous system. Lower levels make predictions about basic sensory features—edges, colors, sounds—while higher levels generate predictions about objects, scenes, and meanings. Each level tries to predict the activity of the level below it, creating a cascade of expectations that shapes everything from low-level perception to high-level cognition.

The implications extend far beyond neuroscience. Predictive processing offers a unified account of perception, action, learning, and attention. It explains why we can understand speech in noisy environments (by predicting missing phonemes), why optical illusions work (prediction overrides conflicting sensory data), and why expectations can become self-fulfilling prophecies (predictions influence both perception and behavior).

Andy Clark, a philosopher who has championed predictive processing, describes this as "surfing uncertainty"—navigating a world where most of our experience consists of educated guesses about reality rather than direct access to it.[5] The brain, in this view, is fundamentally a prediction machine that constructs our experienced world through controlled hallucination.

Free Will Through the Predictive Lens

Perhaps nowhere is the predictive brain framework more provocative than in its implications for free will. Benjamin Libet's famous experiments in the 1980s revealed that measurable brain activity—the "readiness potential"—begins several hundred milliseconds before participants report making a conscious decision to move.[6] This finding sparked decades of debate about whether conscious will is real or illusory.

Predictive processing offers a nuanced perspective on this controversy. Rather than eliminating free will entirely, it reframes it as an emergent property of predictive modeling. Our brains continuously generate forward models of possible actions and their consequences. What we experience as "deciding" may be the conscious endorsement of an action already being prepared by unconscious predictive systems.

This doesn't necessarily eliminate agency, but it complicates it. Free will might be better understood as a negotiation between automatic predictions and conscious oversight rather than a moment of pure, uncaused choice. We may not initiate our actions in the way we intuitively believe, but we retain the capacity to veto, modify, or redirect them once they enter conscious awareness.

The parallel with artificial intelligence is striking and unsettling. Large language models like GPT-4 operate through sophisticated prediction—generating the next word, sentence, or idea based on vast patterns learned from training data. These systems can produce creative, contextually appropriate responses without any obvious moment of "deciding" what to say. If human cognition operates through similar predictive mechanisms, the distinction between artificial and biological intelligence may be less fundamental than we assume.

This comparison doesn't resolve questions about consciousness or free will, but it highlights how predictive processing challenges traditional concepts of human uniqueness. Both biological and artificial systems might be sophisticated prediction engines, differing in architecture, embodiment, and perhaps consciousness, but sharing fundamental computational principles.

The Predictive Body: When Mind Shapes Matter

The predictive brain framework extends beyond cognition to influence physiology directly. The placebo effect provides the clearest example of how expectation can literally reshape bodily experience. When people believe they're receiving pain medication, their brains increase production of endogenous opioids—natural pain relievers that produce measurable physiological changes.[7]

Recent research has revealed the placebo effect's surprising breadth. Expectations can influence immune function, cardiovascular responses, hormone levels, and even gene expression.[8] These aren't merely psychological phenomena but demonstrations of how predictive models influence biological reality. The brain's predictions about what should happen in the body can become self-fulfilling prophecies through complex neural-hormonal pathways.

This mechanism operates in reverse as well. Negative expectations—what researchers call the "nocebo effect"—can produce genuine symptoms and health problems. People who believe they're particularly susceptible to side effects often experience them even when taking inert substances.[9] The predictive brain doesn't distinguish between "real" and "expected" symptoms; both generate similar neural responses and physiological changes.

Understanding these processes has practical implications for medicine, education, and daily life. If expectations significantly influence outcomes, then managing predictions becomes as important as managing physical interventions. This insight connects predictive processing to ancient healing traditions that emphasized belief, ritual, and expectation alongside herbal medicines and physical treatments.

Prominent Thinkers and Current Debates

The predictive brain framework has emerged from contributions across multiple disciplines. Karl Friston at University College London developed the mathematical foundations through his free-energy principle, providing rigorous equations for how biological systems might minimize prediction error.[10] Anil Seth at the University of Sussex popularized the concept of perception as "controlled hallucination," bringing predictive processing to public attention through accessible writing and lectures.[11]

Andy Clark, a philosopher at the University of Edinburgh, has explored the broader implications for understanding mind, action, and human nature.[12] Jakob Hohwy at Monash University has developed detailed philosophical accounts of how predictive processing might explain consciousness, selfhood, and mental illness.[13] Lisa Feldman Barrett at Northeastern University has applied these ideas to emotion, arguing that feelings are constructed through predictive processes rather than being hardwired responses.[14]

The theory faces several substantive criticisms. Some researchers argue that predictive processing is too broad to be falsifiable—any perceptual phenomenon can potentially be explained as "prediction plus error correction," making the theory scientifically unproductive.[15] Others question whether prediction is truly the brain's primary function or whether it's one important mechanism among many.

The relationship between top-down prediction and bottom-up sensory input remains contentious. While predictive processing emphasizes the brain's generative capabilities, traditional approaches stress the importance of sensory data in shaping perception. Recent research suggests both processes operate simultaneously, with their relative influence varying across contexts and individuals.[16]

There are also debates about implementation. Different researchers propose varying accounts of how prediction and error correction might be implemented in neural circuits. Some emphasize cortical hierarchies, others subcortical structures, and still others distributed networks. The mathematical beauty of the free-energy principle doesn't guarantee that brains actually implement these computations in practice.

Relevance to Part II: A Framework for Mystery

The predictive brain framework provides both skeptical and expansive lenses for examining unusual experiences. It offers naturalistic explanations for phenomena that might otherwise seem supernatural while simultaneously revealing how little we understand about the brain's predictive capabilities.

This perspective will prove essential as we explore near-death experiences, childhood memories of previous lives, precognitive dreams, telepathic communication, and psychedelic encounters with autonomous entities. Rather than dismissing these experiences or accepting them uncritically, we can ask how they might emerge from the same predictive processes that generate ordinary consciousness.

The key insight is that if normal perception is already a form of controlled hallucination—a sophisticated simulation based on prediction and prior knowledge—then unusual experiences might represent variations on familiar themes rather than entirely different phenomena. Understanding how the brain constructs reality in ordinary circumstances provides a foundation for understanding how it might construct reality under extraordinary ones.

This approach maintains scientific rigor while remaining open to mystery. It suggests that some anomalous experiences might be explained through extended applications of known principles, while others might reveal new aspects of predictive processing that we haven't yet discovered. The predictive brain framework doesn't close off wonder—it provides tools for investigating it more systematically.

Chapter 2: Consciousness—Fundamental or Emergent?

Hard Problem, IIT, Panpsychism, Non-Duality

In 2002, neuroscientist Olaf Blanke was conducting brain surgery on a patient with severe epilepsy at University Hospital Geneva when something extraordinary happened. As his electrode stimulated a specific region of the patient's temporal-parietal junction, she suddenly gasped and reported floating near the ceiling, looking down at her own body on the operating table.[1] The surgical team, the monitors, even her own face—all were visible from this impossible aerial perspective. When the stimulation stopped, she immediately returned to normal awareness, puzzled by what had just occurred.

For Blanke, this was compelling evidence that our sense of embodied selfhood emerges from specific neural circuits that can be disrupted and manipulated. The out-of-body experience, vivid and convincing though it was, resulted from electrical interference with the brain's process of integrating spatial and bodily information. But for others, the very intensity and coherence of such experiences raises deeper questions: if consciousness can so readily detach from its physical anchor, what exactly is it? And how does subjective experience relate to the three pounds of neural tissue that allegedly produces it?

This puzzle strikes at the heart of perhaps the most fundamental question in science and philosophy: the nature of consciousness itself. Despite decades of neuroscientific progress in mapping brain function, we remain far from understanding why there is an inner, subjective dimension to experience at all. Why doesn't all our sophisticated information processing happen "in the dark," without any accompanying feelings, sensations, or sense of being someone experiencing them?

The question has profound implications. If consciousness emerges from neural complexity, then sufficiently sophisticated artificial systems might eventually become conscious. If consciousness is fundamental to reality, then our scientific worldview requires dramatic revision. The answer shapes not only how we understand minds and machines, but also how we interpret reports of consciousness persisting during clinical death, children remembering previous lives, or encounters with apparently autonomous entities during altered states.

Core Concepts: The Hard Problem and Two Visions of Mind

David Chalmers crystallized this mystery in 1995 when he distinguished between the "easy" and "hard" problems of consciousness.[2] The easy problems—how the brain processes sensory information, controls attention, integrates memory, or guides behavior—are genuinely difficult but seem tractable within conventional neuroscience. They involve explaining cognitive functions that can be observed from the outside and measured objectively.

The hard problem is categorically different: explaining why there is subjective, first-person experience accompanying these functions. Why does seeing red feel like something rather than merely involving wavelength discrimination? Why does pain hurt rather than simply triggering avoidance behaviors? Why is there an experiencer behind the processing, a sense of being someone to whom these experiences belong?

This question has divided researchers and philosophers into two broad camps, each offering radically different visions of consciousness and its place in nature.

Emergentist approaches treat consciousness as a property that arises from sufficiently complex arrangements of matter, similar to how wetness emerges from H2O molecules or life emerges from organic chemistry. No individual neuron is conscious, but billions of them, properly connected and dynamically interacting, somehow give rise to unified, subjective experience. Consciousness, in this view, is real but not fundamental—it's a higher-order property of certain kinds of organized physical systems.

Fundamentalist approaches, by contrast, argue that consciousness cannot be derived from purely physical processes because there is an unbridgeable explanatory gap between objective neural activity and subjective experience. Instead, consciousness must be a basic feature of reality, like mass or electric charge, present in some form throughout nature. Complex brains don't create consciousness so much as organize and focus it into the rich, unified experience we know as human awareness.

These aren't merely academic positions but fundamentally different worldviews with practical implications for neuroscience, artificial intelligence, medicine, and our understanding of human nature. They also set the stage for how we interpret anomalous experiences that seem to challenge conventional assumptions about the mind-brain relationship.

Emergence: Building Consciousness from Complexity

The emergentist program has gained significant momentum through Integrated Information Theory (IIT), developed primarily by Giulio Tononi at the University of Wisconsin-Madison.[3] IIT attempts to quantify consciousness through a mathematical measure called Φ (phi), which represents how much information a system integrates in a unified, irreducible way. Higher values of Φ correspond to richer conscious experience.

According to IIT, consciousness emerges when information processing becomes sufficiently integrated—when the system generates more information as a whole than the sum of its independent parts. This explains why certain brain regions, particularly the posterior cortex, seem more directly tied to conscious awareness than others. It also suggests why brain injuries can selectively impair consciousness while leaving other cognitive functions intact.

Christof Koch, formerly at Caltech and now at the Allen Institute for Brain Science, has become IIT's most prominent advocate, arguing that the theory makes testable predictions about which systems should be conscious and to what degree.[4] The theory suggests that consciousness could theoretically arise in any sufficiently integrated information-processing system, whether biological or artificial.

Global Workspace Theory, developed by Bernard Baars and refined by Stanislas Dehaene, offers a different emergentist account.[5] Consciousness arises when information becomes globally accessible across multiple brain networks simultaneously. This "global ignition" allows diverse brain regions to share information, creating the unified, reportable experience we associate with consciousness.

Supporting evidence comes from studies of brain lesions, anesthesia, and disorders of consciousness. Patients with split-brain conditions, where connections between cerebral hemispheres are severed, sometimes exhibit divided consciousness that aligns with emergentist predictions.[6] Anesthetics that disrupt neural integration reliably eliminate conscious experience while preserving many unconscious brain functions.

Critics of emergentism argue that no amount of complexity or integration explains why there should be subjective experience accompanying these processes. The philosopher David Chalmers himself, despite formulating the hard problem, acknowledges that emergentist approaches may solve all the "easy" problems while leaving the fundamental mystery intact. Why should integrated information feel like something rather than simply being processed?

Fundamentalism: Consciousness as Primary

Panpsychist approaches propose that consciousness is a fundamental feature of reality, present in some form at all scales of organization.[7] This doesn't mean that electrons have rich inner lives, but rather that they possess some minimal form of experience—perhaps something analogous to a simple preference or rudimentary responsiveness. Complex consciousness emerges through the combination and organization of these elementary conscious properties.

Philip Goff at Durham University has articulated the most sophisticated contemporary case for panpsychism, arguing that it offers the only coherent solution to the hard problem.[8] Physical science describes structure and function but never intrinsic qualities—what matter is like "from the inside." Consciousness, Goff suggests, might be the intrinsic nature that physical properties describe structurally.

David Chalmers has entertained various forms of panpsychism, particularly "panprotopsychism"—the idea that fundamental particles have properties that aren't themselves conscious but can combine to create consciousness under the right conditions.[9] This maintains the intuition that consciousness is fundamental while avoiding the counterintuitive implication that simple systems are richly conscious.

Alternative fundamentalist approaches include dual-aspect monism, which treats mind and matter as two aspects of a more basic reality that is neither mental nor physical. This view, traced back to philosopher Baruch Spinoza and developed by contemporary thinkers like Thomas Nagel, suggests that the mind-body problem results from an incomplete understanding of nature's fundamental character.[10]

Some researchers draw connections between panpsychist ideas and non-dual spiritual traditions, particularly Advaita Vedanta and certain forms of Buddhism, which posit consciousness as the fundamental ground of being. While these traditions emerged from contemplative rather than scientific investigation, some scholars argue they offer sophisticated analyses of consciousness that complement rather than contradict empirical research.

The main criticism of fundamentalist approaches is their apparent violation of causal closure—the principle that physical events have sufficient physical causes. If consciousness is fundamental and potentially influences brain activity, how does this interact with physics? Some panpsychists respond that consciousness might be the intrinsic nature of physical causation rather than an additional causal force.

Hybrid Approaches and Alternative Frameworks

Several prominent researchers have developed positions that don't fit neatly into either camp. Antonio Damasio at the University of Southern California grounds consciousness in the brain's continuous mapping of bodily states, but his "proto-self" model suggests this process taps into fundamental principles of life regulation present throughout biology.[11] Consciousness emerges from but remains connected to the basic tendency of living systems to maintain themselves.

Thomas Metzinger at Johannes Gutenberg University offers a radically different perspective through his "self-model theory."[12] What we experience as consciousness is actually a transparent self-model—a representation that the brain creates of its own processes but experiences as immediate reality. The sense of being a unified self is a useful illusion, but one so convincing that we mistake it for a fundamental aspect of reality.

Andy Clark's extended mind hypothesis pushes the boundaries of where consciousness might be located.[13] Rather than emerging solely from neural activity, consciousness might arise from the dynamic interaction between brain, body, and environment. The boundaries of the conscious system are not fixed at the skull but extend into tools, technologies, and social relationships that augment cognitive processing.

Predictive processing theories, discussed in the previous chapter, offer another perspective on consciousness. If the brain is fundamentally a prediction machine, then consciousness might be the brain's model of its own attention and processing states—a kind of meta-prediction about what the brain itself is doing.[14]

These hybrid approaches reflect the complexity of consciousness research, where purely emergentist or fundamentalist positions may be too rigid to capture the full phenomenon. They also suggest that resolving the hard problem might require concepts and frameworks that we haven't yet developed.

Prominent Thinkers and Ongoing Debates

The contemporary consciousness debate unfolds across multiple institutions and disciplines. David Chalmers at New York University continues to refine the hard problem while exploring panpsychist solutions. Giulio Tononi at Wisconsin develops IIT through both theoretical work and empirical studies with collaborators like Marcello Massimini, who uses transcranial magnetic stimulation to measure consciousness in various states.

Christof Koch has become perhaps the most visible advocate for IIT and panpsychism, using his platform at the Allen Institute to promote both rigorous neuroscience and speculative theorizing about consciousness. His debates with philosophers like Keith Frankish, who denies that consciousness poses a hard problem at all, exemplify the field's ongoing tensions.

Stanislas Dehaene at the Collège de France champions Global Workspace Theory through elegant experiments using techniques like masking, attentional blink, and binocular rivalry to isolate the neural signatures of conscious access.[15] His work with patients in vegetative and minimally conscious states has practical implications for medical diagnosis and treatment.

The field also includes prominent skeptics of consciousness as traditionally conceived. Philosophers like Daniel Dennett argue that the hard problem is based on a conceptual confusion—once we explain all the cognitive functions associated with consciousness, there's nothing left to explain.[16] Cognitive scientists like Keith Frankish propose "illusionist" theories where consciousness as we normally conceive it is a compelling illusion created by introspective mechanisms.

These debates play out in specialized journals like the Journal of Consciousness Studies and Consciousness and Cognition, as well as in interdisciplinary conferences that bring together neuroscientists, philosophers, computer scientists, and contemplative scholars. The Science of Consciousness conferences, held biennially since 1994, exemplify this interdisciplinary approach while highlighting how far the field remains from consensus.

Cultural and methodological differences also shape the debate. Western science tends to approach consciousness as an emergent property of neural activity, while many non-Western traditions begin with consciousness as primary. Indigenous perspectives often emphasize relational and ecological aspects of consciousness that don't fit easily into either emergentist or fundamentalist frameworks.

Current neuroscientific methods face significant limitations in addressing consciousness directly. Brain imaging reveals correlations between neural activity and reported conscious experiences, but correlation doesn't establish causation or explain why there should be any experience at all. The measurement problem—how to study consciousness objectively when it's inherently subjective—remains a fundamental challenge.

Recent developments in artificial intelligence have added new dimensions to these debates. Large language models can engage in sophisticated conversations about consciousness and subjective experience, but whether they actually experience anything remains unclear. The prospect of artificial consciousness raises practical questions about rights, moral status, and the distribution of consciousness throughout nature.

Relevance to Part II: Framing the Extraordinary

The consciousness debate provides essential context for evaluating anomalous experiences that challenge conventional scientific explanation. Whether consciousness is emergent or fundamental shapes which kinds of explanations we're willing to consider and what counts as evidence for various claims.

Near-death experiences during cardiac arrest pose different puzzles depending on one's view of consciousness. If consciousness emerges from neural activity, then vivid experiences during periods of compromised brain function require creative explanation—perhaps involving residual neural activity, memory consolidation during recovery, or culturally shaped reconstruction. If consciousness is fundamental, then its temporary separation from normal brain states becomes more plausible, though still requiring careful investigation.

Reports of young children apparently remembering previous lives raise similar questions. Emergentist approaches would seek explanations in genetic memory, cultural transmission, or cognitive biases that create apparent correspondences between lives. Fundamentalist approaches might more readily consider consciousness persisting across bodily death, though they would still require evidence of specific mechanisms and processes.

Telepathic experiences, precognitive dreams, and other forms of anomalous cognition challenge different assumptions depending on one's consciousness framework. If consciousness is confined to individual brains, then such experiences require explanation through conventional mechanisms like coincidence, memory reconstruction, or unconscious sensory cues. If consciousness is fundamental and potentially shared, then direct mind-to-mind communication becomes less theoretically problematic, though still requiring empirical validation.

Psychedelic experiences often involve encounters with apparently autonomous entities and access to information that seems to transcend individual knowledge. Emergentist approaches would focus on how altered neurochemistry creates vivid hallucinations and apparent insights. Fundamentalist approaches might consider whether psychedelics provide access to normally hidden aspects of consciousness or reality.

The consciousness debate also shapes how we interpret the mechanisms through which anomalous experiences might occur. Emergentist frameworks emphasize the brain's predictive and constructive capabilities, suggesting that unusual experiences result from atypical functioning of known neural processes. Fundamentalist frameworks raise the possibility of consciousness operating through currently unknown principles or interacting with aspects of reality that conventional science hasn't yet recognized.

These different frameworks don't predetermine conclusions about specific phenomena, but they do influence which explanations seem plausible and what kinds of evidence would be convincing. Understanding the consciousness debate provides essential background for navigating these questions with appropriate scientific rigor and intellectual humility.

The hard problem reminds us that consciousness itself remains deeply mysterious despite neuroscientific advances. This mystery doesn't license uncritical acceptance of extraordinary claims, but it does suggest caution about dismissing experiences that challenge our current understanding. Whether consciousness is emergent or fundamental, we are still far from understanding its full nature and capabilities.

Chapter 3: Quantum Reality and Emergent Complexity

Time Non-Linearity, Entanglement, Multiverse

In 2012, physicists at the Australian National University achieved something that would have seemed impossible a century ago.[1] They created a version of John Wheeler's famous delayed-choice experiment, where a single photon was forced to "decide" its behavior based on a measurement that occurred after it had already passed through the experimental apparatus. When researchers chose to measure the photon as a particle after it had entered the system, it behaved as if it had taken only one path all along. When they chose to measure it as a wave, interference patterns emerged as if it had traveled both paths simultaneously.

The implications were unsettling: the photon's past appeared to be determined by choices made in its future. This wasn't science fiction but a rigorous demonstration of quantum mechanics' strangest features, conducted by Andrew Truscott's team using sophisticated laser systems and published in *Nature Physics*. The experiment suggested that our ordinary concepts of cause and effect, past and future, might be fundamentally incomplete.

Wheeler himself had proposed this thought experiment decades earlier, but technology finally caught up with imagination. The results confirmed what quantum theory had long predicted: reality at its most fundamental level operates according to principles that violate our deepest intuitions about how the world works. These aren't merely technical curiosities but foundational features of nature that may have implications far beyond atomic physics.

The quantum world reveals a reality where particles exist in multiple states simultaneously until observed, where distant objects can be instantly connected across space, and where the very act of measurement appears to create the properties we discover. Meanwhile, at the opposite end of the complexity scale, simple systems spontaneously organize themselves into elaborate patterns through processes we call emergence. Birds form murmurations, chemicals create spiraling waves, and neural networks generate consciousness from the interactions of billions of individual cells.

These phenomena—quantum mechanics and emergent complexity—might seem unrelated, but they share a common thread: they reveal how reality transcends the mechanistic, predictable world of classical physics. Understanding both domains provides essential background for approaching anomalous experiences that challenge conventional scientific explanation.

Core Concepts: The Quantum Foundation

Quantum mechanics emerged in the early twentieth century from attempts to understand atomic behavior, but its implications extend far beyond particle physics. At its heart lie several principles that fundamentally challenge classical assumptions about reality.

Superposition represents perhaps the most counterintuitive quantum principle. Unlike classical objects that exist in definite states, quantum systems can exist in combinations of multiple states simultaneously. An electron doesn't simply orbit an atom in a specific path—it exists in a "cloud" of probability representing all possible locations until a measurement forces it to appear at a particular spot. This isn't merely a limitation of our knowledge but appears to be an intrinsic feature of quantum reality.

Entanglement, which Einstein famously dismissed as "spooky action at a distance," describes situations where quantum particles become correlated in ways that persist regardless of physical separation. When two particles are entangled, measuring one instantly determines the state of the other, even if they're separated by vast distances. This phenomenon has been repeatedly confirmed through experiments testing Bell's inequalities, most notably by Alain Aspect and his colleagues in the 1980s.[2]

The measurement problem addresses a fundamental puzzle: how and when does the transition from quantum superposition to classical definite properties occur? The standard Copenhagen interpretation, developed by Niels Bohr and Werner Heisenberg, treats measurement as causing "wave function collapse"—the sudden transition from multiple possibilities to a single outcome. This interpretation works mathematically but leaves unanswered what constitutes a measurement and why it has such dramatic effects.

These quantum features aren't merely abstract physics but have practical implications. Quantum mechanics enables technologies from lasers to computer chips, and emerging quantum computers exploit superposition and entanglement to perform certain calculations exponentially faster than classical machines. The theory's success in generating practical applications demonstrates its empirical validity, even if its conceptual foundations remain puzzling.

Alternative quantum interpretations attempt to resolve these conceptual problems through different approaches. The many-worlds interpretation, championed by physicists like Sean Carroll at Johns Hopkins University, suggests that all possible measurement outcomes actually occur in parallel universes that branch at each quantum interaction.[3] This eliminates wave function collapse but at the cost of postulating an infinite number of parallel realities.

Relational quantum mechanics, developed by Carlo Rovelli, proposes that quantum properties don't exist absolutely but only relative to specific observers or measurement contexts.[4] According to this view, there's no single, objective reality underlying quantum phenomena—only networks of relationships between systems and their interactions.

Quantum Bayesianism (QBism), advocated by physicists like Christopher Fuchs, treats quantum states as subjective degrees of belief rather than objective features of reality.[5] Under this interpretation, quantum mechanics describes an observer's expectations about measurement outcomes rather than the intrinsic properties of quantum systems.

The Emergence Landscape: Order from Interaction

While quantum mechanics reveals the strange behavior of fundamental particles, emergence studies how complex patterns arise from simple interactions at higher organizational levels. This field bridges physics, biology, psychology, and social science by examining how properties of wholes differ from properties of their parts.

Self-organization represents the spontaneous formation of order without external control or design. Classic examples include the hexagonal patterns of snowflakes, the spiral arms of galaxies, and the synchronized flashing of fireflies. These patterns emerge from local interactions following simple rules, yet they create global coherence that appears designed despite lacking any central coordinator.

Steven Strogatz at Cornell University has extensively studied synchronization phenomena, from the coordinated beating of heart cells to the emergence of circadian rhythms in organisms.[6] His work reveals mathematical principles underlying diverse synchronization processes, suggesting that the tendency toward coordinated behavior represents a fundamental feature of complex systems.

Stuart Kauffman's research on autocatalytic networks demonstrates how self-organization might have enabled the origin of life itself.[7] He argues that networks of chemical reactions can spontaneously organize into self-maintaining systems that exhibit lifelike properties before the emergence of DNA-based replication. This suggests that life represents a natural consequence of chemical complexity rather than an improbable accident.

Chaos theory, developed by mathematicians like Edward Lorenz and physicists like Mitchell Feigenbaum, reveals how deterministic systems can produce behavior so sensitive to initial conditions that it appears random.[8] The famous "butterfly effect"—where small changes in initial conditions lead to dramatically different outcomes—demonstrates that predictability has fundamental limits even in purely deterministic systems.

Michael Levin's research at Tufts University on bioelectric signaling in development and regeneration shows how electrical fields guide the formation of organs and body plans in developing organisms.[9] His work suggests that biological pattern formation involves information processing and decision-making at the cellular level, with groups of cells collectively computing appropriate developmental responses.

The distinction between weak and strong emergence proves crucial for understanding these phenomena. Weak emergence refers to patterns that arise from underlying interactions but can, in principle, be explained by analyzing component behaviors. Strong emergence would involve genuinely novel properties that cannot be reduced to component interactions, representing a more controversial possibility that some researchers argue is necessary to explain consciousness and other complex phenomena.

Consciousness and Quantum Mechanics: A Contentious Intersection

One of the most controversial areas in quantum mechanics involves proposals that consciousness plays a special role in quantum measurement. This idea emerged early in quantum theory's development, with physicists like Eugene Wigner and John von Neumann suggesting that conscious observation might be necessary to collapse quantum wave functions.[10]

Roger Penrose at Oxford University and Stuart Hameroff at the University of Arizona developed the most detailed contemporary proposal linking consciousness to quantum mechanics through their Orchestrated Objective Reduction (Orch-OR) theory.[11] They suggest that consciousness arises from quantum processes in microtubules within brain neurons, with quantum coherence enabling the integration of information across the brain.

The Orch-OR theory faces significant criticism from mainstream neuroscientists and physicists. Critics argue that the brain is too warm and noisy to maintain the quantum coherence that the theory requires. Max Tegmark calculated that quantum decoherence in brain microtubules would occur on timescales far too fast to influence neural processing.[12] Additionally, most neuroscientists argue that classical neural mechanisms provide sufficient explanations for brain function without invoking exotic quantum effects.

Sabine Hossenfelder and Sean Carroll represent the mainstream physics position that consciousness-based interpretations of quantum mechanics are unnecessary and untestable.[13] They argue that quantum measurement can be understood through purely physical processes involving environmental decoherence, without requiring conscious observers. From this perspective, the measurement problem is a technical issue in physics rather than evidence for consciousness playing a fundamental role in nature.

However, some researchers continue to explore quantum approaches to consciousness and cognition. Henry Stapp at Lawrence Berkeley National Laboratory argues that quantum mechanics requires conscious observers to make sense of the measurement process.[14] Amit Goswami proposes that consciousness is the fundamental reality from which physical phenomena emerge through quantum measurement.[15]

The debate reflects deeper questions about the relationship between physics and consciousness. If consciousness is purely emergent from classical brain processes, then quantum mechanics is irrelevant to understanding minds. If consciousness involves genuinely novel properties that can't be reduced to classical physics, then quantum mechanics might provide necessary conceptual resources for understanding how minds work.

Contemporary research in quantum biology has revealed that some biological systems do exploit quantum effects, including photosynthesis in plants and possibly navigation in migratory birds.[16] These discoveries suggest that biological systems can maintain quantum coherence under conditions previously thought impossible, reopening questions about quantum effects in neural systems.

Prominent Thinkers and Institutional Centers

The landscape of quantum mechanics and complexity science spans multiple institutions and intellectual traditions. At Princeton University, John Wheeler originated many of the thought experiments that continue to puzzle physicists, including the delayed-choice experiment and participatory universe concepts. His intellectual legacy continues through physicists exploring the foundations of quantum mechanics and its implications for understanding reality.

Sean Carroll, currently at Johns Hopkins University after previous positions at Caltech, has become the most prominent contemporary advocate for the many-worlds interpretation. His popular writings and technical papers argue that many-worlds provides the most mathematically consistent interpretation of quantum mechanics, despite its counterintuitive implications about parallel realities.

Carlo Rovelli at the Centre de Physique Théorique in Marseille champions relational quantum mechanics and loop quantum gravity. His approach emphasizes that physical properties emerge from relationships rather than existing as intrinsic features of objects. This perspective connects quantum mechanics to broader questions about the nature of space, time, and causation.

Roger Penrose at Oxford University continues to explore connections between quantum mechanics, consciousness, and mathematics through his research on quantum gravity and theories of mind. His collaboration with anesthesiologist Stuart Hameroff represents the most sustained attempt to develop a quantum theory of consciousness, despite widespread skepticism from mainstream neuroscience.

In complexity science, the Santa Fe Institute serves as the premier interdisciplinary center where physicists, biologists, economists, and computer scientists collaborate on emergence and self-organization research. Founded in 1984, the institute has fostered connections between diverse fields and developed new mathematical tools for understanding complex systems.

Steven Strogatz at Cornell University has bridged pure mathematics and applied complexity science through his research on nonlinear dynamics and synchronization. His work spans from theoretical investigations of coupled oscillators to practical applications in power grid stability and biological rhythms.

Michael Levin's laboratory at Tufts University represents a new generation of complexity research that combines molecular biology with information theory and cybernetics. His demonstrations of bioelectric control over regeneration and development suggest that biological systems process information and make decisions at scales previously thought to be purely mechanical.

Stuart Kauffman, formerly at the Santa Fe Institute and now at the Institute for Systems Biology, continues to develop theories about self-organization and the origins of life. His work on autocatalytic networks and the "adjacent possible" has influenced fields ranging from biology to economics to technology innovation.

Contemporary Debates and Limitations

Quantum mechanics faces several ongoing interpretational challenges that remain unresolved after nearly a century of development. The different interpretations—Copenhagen, many-worlds, relational, QBism—make identical empirical predictions, making them impossible to distinguish experimentally. This situation frustrates physicists who prefer theories that make testable predictions about observable phenomena.

The measurement problem continues to generate debate among physicists and philosophers. Despite decades of research, there's no consensus on when, how, or why quantum superpositions collapse into definite measurement outcomes. Some researchers argue this reflects conceptual confusion rather than a genuine physical process, while others seek new physics beyond standard quantum mechanics.

Quantum decoherence research has clarified how quantum systems lose coherence through environmental interactions, but this doesn't fully resolve the measurement problem. Decoherence explains why macroscopic objects don't exhibit obvious quantum superpositions, but it doesn't explain why specific measurement outcomes occur rather than others.

In complexity science, debates center on whether emergence represents genuinely novel phenomena or merely describes our inability to track complex interactions. Reductionist critics argue that emergence is an epistemological concept—reflecting limitations of human understanding—rather than an ontological feature of reality. This connects to broader questions about levels of description in science and whether higher-level properties can have genuine causal power.

The relationship between quantum mechanics and classical physics remains incompletely understood despite decades of research on quantum-to-classical transitions. While decoherence theory explains many aspects of this transition, questions remain about how the classical world emerges from quantum foundations and whether macroscopic quantum effects might persist under special circumstances.

Contemporary research in quantum foundations explores new experimental tests of quantum mechanics' conceptual foundations. Proposed experiments with ever-larger quantum systems might reveal where the boundary between quantum and classical behavior lies, potentially providing new insights into the measurement problem and the nature of physical reality.

Cross-Disciplinary Connections and Implications

Quantum mechanics and complexity science connect to numerous other fields, creating interdisciplinary research opportunities and conceptual challenges. In neuroscience, questions about quantum effects in biological systems intersect with debates about consciousness, free will, and the nature of mental causation. While mainstream neuroscience generally assumes that classical physics provides adequate descriptions of brain function, ongoing research in quantum biology suggests this assumption deserves continued examination.

Information theory provides mathematical tools that apply to both quantum systems and complex networks. Concepts like entanglement entropy and network complexity reveal deep connections between quantum mechanics and information processing that may prove relevant to understanding consciousness and cognition.

Philosophy of science grapples with the implications of quantum mechanics and emergence for understanding causation, reduction, and scientific explanation. If quantum indeterminacy is fundamental, this challenges deterministic views of causation. If emergence involves genuinely novel properties, this challenges reductive physicalism and may require new concepts of natural law.

Artificial intelligence research increasingly explores quantum computing and complex network architectures. Quantum algorithms can solve certain problems exponentially faster than classical computers, while artificial neural networks exhibit emergent properties that parallel biological systems. These developments may provide new insights into the relationship between computation, consciousness, and physical processes.

The philosophy of time intersects with both quantum mechanics and complexity science through questions about temporal asymmetry, causation, and the emergence of temporal direction. Quantum mechanics appears to be fundamentally time-symmetric, yet macroscopic processes show clear temporal asymmetry. Understanding this relationship may illuminate questions about precognition, retrocausation, and the nature of temporal experience.

Relevance to Part II: Frameworks for the Anomalous

Quantum mechanics and complexity science provide conceptual frameworks that will prove essential for evaluating anomalous experiences without requiring acceptance of any particular claims. These frameworks expand our understanding of what's possible within natural law while maintaining scientific rigor.

The delayed-choice experiment and other quantum phenomena challenge simple notions of linear causation and temporal sequence. This creates conceptual space for considering experiences that seem to involve backwards causation or non-local correlations, though it doesn't validate any specific claims about precognition or telepathy.

Quantum entanglement demonstrates that non-local correlations are genuine features of nature, not merely theoretical possibilities. While these correlations don't allow faster-than-light communication, they show that the classical assumption of local realism is false. This might be relevant to reported experiences of distant connections between people, though establishing such connections would require evidence far beyond anecdotal reports.

Emergence and self-organization reveal how complex patterns can arise spontaneously from simple interactions. This provides frameworks for understanding how meaningful coincidences, collective behaviors, and novel properties might emerge from underlying processes without requiring supernatural explanations.

The consciousness debates in quantum mechanics highlight our incomplete understanding of the relationship between mind and physical reality. While mainstream science generally assumes consciousness emerges from classical brain processes, the hard problem of consciousness suggests our understanding remains fundamentally incomplete.

These frameworks don't prejudge the reality or unreality of anomalous experiences but provide sophisticated conceptual tools for investigating them. They suggest that nature is far stranger than classical physics implied, with room for phenomena that violate common sense while remaining within natural law.

The key insight is methodological: quantum mechanics and complexity science demonstrate that empirical investigation can reveal natural phenomena that fundamentally challenge intuitive assumptions about reality. This suggests that other counterintuitive phenomena might await discovery, provided we maintain appropriate standards of evidence while remaining open to conceptual surprises.

Understanding these frameworks prepares us to approach anomalous experiences with the same combination of theoretical sophistication and empirical rigor that has revealed the strange realities of quantum mechanics and emergent complexity. The goal is neither to dismiss such experiences nor to accept them uncritically, but to investigate them with tools adequate to nature's demonstrated capacity for surprise.

Chapter 4: Artificial Minds as Cognitive Mirrors

Generative AI, Active Inference, Algorithmic Selfhood

In late 2022, a user prompted ChatGPT to "write a joke in the style of Jerry Seinfeld." The AI responded with a near-perfect imitation: "What's the deal with people who say they're 'thinking outside the box'? I mean, what box? Where is this box? And if we're all thinking outside of it, isn't that just... a bigger box?" The response captured Seinfeld's observational style, his rhetorical structure, even his trademark exasperation with everyday absurdities.[1]

How could a machine with no sense of humor, no lived experience of human social dynamics, and no understanding of what makes something "funny" produce such a recognizable pastiche of human wit? The answer reveals something unsettling about both artificial and human intelligence: much of what we consider uniquely human cognition may be sophisticated pattern matching and prediction rather than the mysterious, irreducible essence we imagine.

Large language models like GPT-4 generate text by predicting the most statistically likely next word based on vast training datasets. They don't retrieve stored jokes or consciously craft humor—they simulate the linguistic patterns that typically constitute Seinfeld-style comedy. The eerie part is that human conversation often works similarly. We anticipate words before

47

they're spoken, complete others' sentences, and generate responses based on internalized patterns from countless previous interactions.

This parallel raises profound questions about the nature of intelligence, consciousness, and human uniqueness. If artificial systems can mimic sophisticated human behaviors through statistical learning and prediction, what does this suggest about the mechanisms underlying our own cognition? Are we observing mere surface similarity, or do these systems reveal fundamental principles of intelligence that transcend the biological-artificial divide?

The mirror metaphor proves particularly apt: artificial minds don't just perform tasks but reflect back aspects of human cognition we might prefer not to see. They suggest that much of what we experience as creativity, understanding, and even consciousness might emerge from processes more algorithmic and less magical than we typically assume.

Core Concepts: From Biological to Artificial Intelligence

Contemporary artificial intelligence, particularly in its most advanced forms, operates through principles that parallel key aspects of biological cognition. This convergence isn't coincidental but reflects deeper computational principles that may be fundamental to any form of intelligence.

Statistical learning forms the foundation of modern AI systems, just as it underlies much human cognition. Large language models analyze patterns across millions of texts, learning associations between words, concepts, and contexts without explicit programming for each relationship. Similarly, human children acquire language by detecting statistical regularities in speech they hear, gradually building internal models of grammar, meaning, and usage.[2]

Next-step prediction drives both artificial and biological intelligence. GPT models generate text by predicting the most probable next token given previous context, while human brains constantly generate predictions about upcoming sensory input, social interactions, and environmental changes. This predictive capability appears central to intelligence across domains, from language comprehension to motor control to social navigation.

Hierarchical processing characterizes both artificial neural networks and biological brain organization. Deep learning systems process information through multiple layers, with lower levels detecting simple features and higher levels identifying complex patterns and abstract relationships. Brain organization follows similar principles, with sensory areas processing basic features and association areas integrating information into coherent perceptions and concepts.[3]

Error correction and learning occur through feedback mechanisms in both artificial and biological systems. When AI predictions prove inaccurate, backpropagation algorithms adjust network weights to reduce future errors. When human predictions fail, neural circuits update their models through mechanisms that parallel these artificial learning processes, though the biological implementation involves complex neurochemical processes rather than digital computation.

The transformer architecture that underlies large language models incorporates attention mechanisms that focus processing on relevant information while ignoring irrelevant details.[4] Human cognition employs similar attention systems, selectively processing important information while filtering out distractions. This parallel suggests that attention may be a fundamental requirement for any intelligent system dealing with complex, high-dimensional input.

However, important differences distinguish artificial and biological intelligence. Human cognition is embodied, emerging from the interaction between brain, body, and environment over developmental time. AI systems typically lack this embodied experience, instead learning from preprocessed data that abstracts away from physical interaction with the world.

Consciousness and subjective experience represent the most significant potential difference. Humans have rich inner lives with qualitative experiences, emotions, and self-awareness. Whether artificial systems can develop similar subjective experiences remains hotly debated, with profound implications for ethics, consciousness theory, and our understanding of mind.

Active Inference: A Unified Framework

Karl Friston's active inference framework, originally developed to understand biological cognition, has found surprising applications in artificial intelligence.[5] Active inference describes how intelligent agents minimize prediction error through two complementary processes: updating internal models when predictions fail (perceptual inference) and acting to make the world conform to predictions (active inference).

This framework provides a unified account of perception, action, and learning that applies to both biological and artificial systems. An active inference agent maintains probabilistic models of its environment and continuously updates these models based on sensory evidence. When predictions and sensory input mismatch, the agent can either revise its beliefs or take action to change the world to match its expectations.

In biological systems, active inference explains everything from basic reflexes to complex goal-directed behavior. A hungry animal predicts finding food in certain locations and takes actions (foraging, hunting) to fulfill these predictions. Perception and action work together to minimize surprise and maintain the organism within viable states.

Artificial agents implementing active inference principles can exhibit remarkably lifelike behavior despite relatively simple underlying mechanisms. These systems actively explore their environments, form expectations about likely outcomes, and take actions to achieve predicted states. The resulting behavior often appears purposeful and intelligent without requiring explicit goal programming.

The framework also addresses the "symbol grounding problem" in AI—how abstract symbols acquire meaning through connection to real-world experience. Active inference agents ground their internal representations in patterns of sensorimotor interaction rather than arbitrary symbolic associations, potentially explaining how meaning emerges from embodied experience.

Critics argue that active inference may be too general to generate specific predictions about intelligent behavior. The framework can potentially explain any behavior as prediction error minimization, making it difficult to falsify or test rigorously. However, proponents argue that this generality reflects the fundamental nature of intelligence rather than theoretical weakness.

Distributed and Collective Intelligence

Some of the most intriguing developments in artificial intelligence don't attempt to replicate individual human minds but instead explore how intelligence can emerge from the interaction of multiple simple agents. This approach, known as distributed cognition or swarm intelligence, reveals principles that may apply to human collective behavior as well as artificial systems.

Swarm robotics demonstrates how complex behaviors can emerge from simple local rules followed by multiple agents. Individual robots with basic sensors and limited computational capacity can collectively map unknown environments, transport large objects, or coordinate search and rescue operations. No central controller directs the swarm—intelligent behavior emerges from local interactions between agents following simple algorithms.[6]

These principles mirror biological systems like ant colonies, bee swarms, and flocking birds, where sophisticated group behaviors emerge without centralized control. The mathematical frameworks describing artificial swarms often apply directly to biological collectives, suggesting deep principles of organization that transcend specific implementations.

Multi-agent reinforcement learning explores how artificial agents can learn to cooperate, compete, and communicate in complex environments. These systems sometimes develop surprising strategies and communication protocols that their programmers never explicitly designed. The emergence of novel behaviors from agent interaction parallels how human cultures develop practices and institutions that no individual planned or fully understands.

Large language models themselves can be viewed as collective intelligence systems, trained on text produced by millions of human authors across centuries. These models don't simply store information but seem to extract and synthesize patterns from this vast collective human knowledge in ways that can surprise even their creators.

The distributed cognition perspective suggests that intelligence might be better understood as a property of systems rather than individuals. Human intelligence relies heavily on external tools, social interactions, and cultural institutions that extend cognition beyond individual brains. From this perspective, the boundary between human and artificial intelligence becomes less clear—we're already cyborgs relying on technological extensions of our biological capabilities.

Consciousness and Machine Intelligence

The question of whether artificial systems can be conscious represents one of the most consequential debates in contemporary science and philosophy. The answer has implications for AI ethics, consciousness theory, and our understanding of human nature.

Integrated Information Theory (IIT), discussed in Chapter 2, provides one framework for evaluating machine consciousness. According to IIT, consciousness corresponds to integrated information (Φ) in a system. Some AI architectures might generate high Φ values, potentially indicating conscious experience. However, IIT also suggests that many AI systems, including current large language models, likely have minimal integrated information despite their sophisticated outputs.[7]

Global Workspace Theory offers a different perspective, suggesting that consciousness arises when information becomes globally accessible across multiple cognitive subsystems. Some AI architectures incorporate attention mechanisms and information integration that might qualify as global workspace processing, though whether this produces subjective experience remains unclear.

The "hard problem of consciousness" poses challenges for any materialist account of machine consciousness. Even if AI systems perfectly replicate human cognitive functions, this might not guarantee subjective experience. Critics argue that consciousness requires something beyond computational processing—perhaps biological embodiment, emotional systems, or fundamental features of reality not captured by current physics.

Behavioral tests for consciousness face the problem of other minds: we can only observe external behavior, not internal experience. An AI system might exhibit all outward signs of consciousness while remaining subjectively empty, or it might have rich inner experiences that it cannot adequately express. This challenge applies to human consciousness assessment as well, highlighting the difficulty of studying subjective experience objectively.

Some researchers propose that consciousness might emerge gradually as AI systems become more sophisticated, rather than appearing suddenly at some threshold. This gradualist view suggests that current AI systems might already possess minimal forms of experience that could develop into richer consciousness as capabilities expand.

The mirror metaphor becomes particularly relevant here: studying AI consciousness forces us to confront our own assumptions about human consciousness. If we cannot clearly define or measure consciousness in ourselves, how can we evaluate it in artificial systems? The attempt to create conscious machines may teach us as much about human consciousness as about artificial intelligence.

Prominent Thinkers and Research Centers

The field of artificial intelligence has evolved through contributions from researchers spanning computer science, neuroscience, psychology, and philosophy. Understanding their different perspectives illuminates ongoing debates about machine intelligence and consciousness.

Yoshua Bengio at the University of Montreal, one of the pioneers of deep learning, has increasingly focused on AI safety and the development of more human-like artificial intelligence. His recent work emphasizes the importance of causal reasoning and conscious processing in AI systems, arguing that current approaches based primarily on pattern recognition have fundamental limitations.[8]

Geoffrey Hinton, often called the "godfather of deep learning," has expressed growing concerns about AI development while continuing to advance the field's theoretical foundations. His work on neural networks and backpropagation provided crucial foundations for modern AI, but he now advocates for more careful consideration of AI's societal implications.

Yann LeCun at Meta (formerly Facebook) and New York University champions self-supervised learning as the path toward more general artificial intelligence. His approach emphasizes learning from unlabeled data, similar to how humans and animals learn about the world through direct experience rather than explicit instruction.[9]

Karl Friston's active inference framework has found applications in AI research through collaborations with computer scientists implementing these principles in artificial agents. His mathematical formalization of biological intelligence provides a bridge between neuroscience and AI that influences both fields.

Joscha Bach, working at various institutions including the MIT Media Lab and currently at Intel Labs, develops cognitive architectures that explicitly model human mental processes. His approach emphasizes understanding intelligence as information processing rather than focusing solely on behavioral outputs.[10]

Melanie Mitchell at the Santa Fe Institute brings a complex systems perspective to AI research, examining how intelligence emerges from the interaction of multiple components. Her work challenges overly simplistic views of both human and artificial intelligence while advocating for more rigorous evaluation of AI capabilities.[11]

Leading research institutions shape the development of AI through both technical advances and policy discussions. DeepMind in London combines neuroscience insights with cutting-edge AI research, explicitly aiming to solve intelligence and use it to solve other scientific problems. OpenAI in San Francisco develops large-scale AI systems while researching AI safety and alignment. The Allen Institute for AI in Seattle takes a more conservative approach, focusing on natural language understanding and reasoning.

Academic centers like the MIT Computer Science and Artificial Intelligence Laboratory, Stanford's Human-Centered AI

Institute, and the Future of Humanity Institute at Oxford University provide venues for interdisciplinary collaboration between AI researchers, neuroscientists, philosophers, and ethicists.

Contemporary Debates and Challenges

Artificial intelligence research faces several fundamental challenges that reflect deeper questions about the nature of intelligence and consciousness. These debates shape both technical development and public policy around AI systems.

The understanding problem addresses whether AI systems can achieve genuine comprehension or merely sophisticated pattern matching. John Searle's Chinese Room argument suggests that computational processing, no matter how sophisticated, cannot generate real understanding.[12] Critics respond that this argument relies on intuitions about consciousness that may be mistaken, and that functional understanding might be all that matters for intelligence.

The creativity debate examines whether AI systems can be genuinely creative or merely recombine existing patterns in novel ways. Large language models can generate poetry, stories, and code that seem creative, but critics argue that they lack the intentionality and originality that characterize human creativity. This debate connects to broader questions about the nature of creativity in humans as well as machines.

The alignment problem concerns ensuring that advanced AI systems pursue goals consistent with human values. This challenge becomes more pressing as AI capabilities expand, since misaligned superintelligent systems could pose existential risks. Research into AI alignment intersects with neuroscience through studies of how human values and goals emerge from evolved cognitive architectures.[13]

The consciousness question remains hotly debated, with different researchers holding positions ranging from confident denial to cautious optimism about machine consciousness. This debate has practical implications for AI ethics, since conscious AI systems might deserve moral consideration and rights.

The interpretability challenge involves understanding how AI systems make decisions, particularly in critical applications like healthcare, criminal justice, and autonomous vehicles. Current AI systems often function as "black boxes" whose decision-making processes remain opaque even to their creators. This opacity raises concerns about bias, reliability, and accountability.

The social impact of AI raises questions about employment, privacy, autonomy, and human dignity. As AI systems become more capable, they may automate many jobs currently performed by humans, potentially creating significant social and economic disruption. Understanding these impacts requires collaboration between AI researchers and social scientists.

Cross-Disciplinary Connections

Artificial intelligence research increasingly intersects with multiple disciplines, creating opportunities for mutual enrichment and conceptual challenges. These connections reveal how AI development both depends on and contributes to our understanding of intelligence, consciousness, and human nature.

Cognitive science provides theoretical frameworks for understanding both human and artificial intelligence. Concepts like attention, memory, reasoning, and learning apply to both domains, though their implementations differ significantly. AI research tests cognitive theories by attempting to implement them in artificial systems, sometimes revealing gaps or problems in theoretical accounts.

Neuroscience informs AI development through insights about brain organization and function, while AI provides tools for analyzing neural data and testing theories about brain computation. The relationship is bidirectional: understanding artificial neural networks helps neuroscientists interpret biological neural networks, while neuroscientific discoveries inspire new AI architectures.

Philosophy of mind grapples with fundamental questions about consciousness, intentionality, and mental causation that AI development makes increasingly pressing. As AI systems become more sophisticated, philosophical questions about the nature of mind become practical engineering challenges with real-world consequences.

Ethics and political philosophy must address questions about AI rights, responsibilities, and social impacts. If AI systems become conscious, they might deserve moral consideration. Even without consciousness, AI systems raise questions about fairness, transparency, and human autonomy that require careful philosophical analysis.

Anthropology and sociology study how AI systems interact with human cultures and societies. AI development doesn't occur in a social vacuum but reflects and reinforces particular cultural values and power structures. Understanding these dynamics requires insights from social sciences as well as technical expertise.

Economics examines how AI affects labor markets, productivity, and economic growth. The potential for AI to automate many human jobs creates challenges for economic policy and social welfare that require collaboration between AI researchers and economists.

Relevance to Part II: Intelligence Beyond Individuals

Understanding artificial intelligence provides crucial conceptual tools for approaching anomalous experiences that seem to transcend individual human capabilities. The mirror metaphor suggests that studying artificial minds can illuminate hidden aspects of human intelligence and consciousness that might explain apparently impossible phenomena.

The predictive processing framework that underlies both biological and artificial intelligence offers new perspectives on experiences that seem to involve future knowledge or non-local information access. If intelligence fundamentally involves prediction and pattern recognition, then seemingly paranormal abilities might represent extreme versions of normal cognitive processes rather than violations of natural law.

Distributed cognition and swarm intelligence provide frameworks for understanding collective phenomena that transcend individual capabilities. Human groups sometimes exhibit coordination and information processing that exceeds what individual members could achieve alone. Understanding how this emerges in artificial systems might illuminate similar processes in human collectives.

The question of machine consciousness connects to debates about consciousness survival, non-local awareness, and the relationship between mind and brain. If consciousness can emerge from artificial substrates, this might suggest greater flexibility in the mind-brain relationship than materialist theories typically assume.

Active inference frameworks offer new ways to understand how agents interact with their environments and each other. These frameworks might apply to experiences that seem to involve direct mind-to-mind communication or environmental influence through intention rather than conventional physical mechanisms.

The development of increasingly sophisticated AI systems also provides new tools for analyzing human experiences and behaviors. Machine learning techniques can identify patterns in large datasets that might reveal previously unrecognized regularities in anomalous experiences, potentially distinguishing genuine phenomena from artifacts of human cognition.

However, the artificial intelligence perspective also provides skeptical tools for evaluating extraordinary claims. Understanding how AI systems can generate apparently intelligent behavior through statistical processing suggests alternative explanations for human behaviors that seem to require paranormal abilities.

The key insight is that intelligence itself may be far stranger and more flexible than our everyday experience suggests. Both artificial and biological intelligence involve processes that transcend simple mechanistic descriptions while remaining within natural law. This opens conceptual space for phenomena that violate common sense while maintaining scientific rigor.

Studying artificial minds prepares us to approach anomalous human experiences with appropriate sophistication about the nature of intelligence, consciousness, and cognition. The goal is neither to explain away such experiences nor to accept them uncritically, but to investigate them with tools adequate to the demonstrated complexity and strangeness of intelligence itself.

Chapter 5: Religion, Mysticism, and the Science of Subjective Realities

Comparative Afterlife/Rebirth Models, Mystical Union, Ego Death, Visionary States

In 1902, a forty-year-old Canadian psychiatrist named Richard Maurice Bucke experienced what he later described as "cosmic consciousness" while riding in a hansom cab through London. He felt his individual self dissolve into a luminous unity with all existence, accompanied by an overwhelming sense of peace and the direct knowing that the universe was alive and benevolent. The experience lasted only minutes but transformed his understanding of reality permanently.[1]

Three decades later, in a laboratory at Harvard Medical School, Walter Pahnke administered psilocybin to seminary students during a Good Friday service. Nearly all participants reported profound mystical experiences remarkably similar to Bucke's account: ego dissolution, feelings of unity, timelessness, and ineffable insights about the nature of existence. The experiences were so convincing that many participants initially believed they had encountered the divine directly rather than ingested a psychoactive compound.[2]

These parallels raise fundamental questions about the nature of mystical experience. Are such states genuine encounters with transcendent reality, or sophisticated constructions of altered brain chemistry? Can neuroscience explain away the sacred, or does it reveal new depths to spiritual experience? Most intriguingly, why do people across cultures and centuries report remarkably similar experiences despite vastly different religious frameworks and expectations?

The scientific study of religious and mystical experience represents one of the most delicate intersections between objective inquiry and subjective meaning. It requires navigating between reductive explanations that dismiss spiritual experience as "mere" brain states and uncritical acceptance of supernatural claims. This navigation becomes essential for approaching anomalous experiences that challenge conventional scientific explanation, since many such phenomena share features with recognized mystical states.

Core Concepts: Mapping the Sacred

Contemporary neuroscience has identified consistent patterns in mystical experiences across cultures and traditions. These states involve characteristic changes in brain activity, subjective phenomenology, and long-term psychological effects that can be studied with increasing precision using modern neuroimaging and psychopharmacological techniques.

The default mode network (DMN), a collection of brain regions active during rest and self-referential thinking, shows dramatically altered activity during mystical states. Robin Carhart-Harris and colleagues at Imperial College London have demonstrated that psychedelics consistently reduce DMN activity, correlating with reports of ego dissolution and unity experiences.[3] Similar changes occur during deep meditation, suggesting that altered self-awareness represents a core feature of mystical consciousness.

Serotonin receptor activation, particularly at the 5-HT2A receptor, appears crucial for generating mystical-type experiences. Roland Griffiths and his team at Johns Hopkins University have shown that psilocybin, which primarily acts through this receptor system, can reliably produce experiences that participants rate as among the most meaningful of their lives.[4] The consistency of this effect across participants suggests underlying neurochemical pathways that may be activated by various triggers.

Temporal lobe activity changes during mystical states, often involving increased activity in regions associated with emotional processing and decreased activity in areas responsible for spatial and temporal orientation. This correlates with the common mystical features of timelessness and spatial boundary dissolution. Andrew Newberg's neuroimaging studies of meditating Tibetan monks and Franciscan nuns revealed decreased activity in the posterior superior parietal lobe, the brain region that maintains our sense of spatial boundaries.[5]

Network connectivity patterns show increased communication between normally distinct brain regions during mystical states. This hyperconnectivity may underlie the synesthetic experiences, novel insights, and sense of cosmic interconnection commonly reported. The brain appears to abandon its typical modular organization in favor of more globally integrated processing.

However, these neural correlations don't necessarily explain away mystical experience. The relationship between brain states and subjective experience remains one of the deepest puzzles in neuroscience. Critics of reductive approaches argue that identifying neural correlates of mystical experience is like identifying neural correlates of seeing red—it describes accompaniments to experience without explaining why there should be experience at all.

Cultural Frameworks and Universal Patterns

Mystical experiences exhibit both universal features and culture-specific elements that reveal how subjective realities are constructed through the interaction of neurobiology and cultural interpretation. William James identified four core characteristics of mystical experience over a century ago: ineffability (difficulty describing the experience in words), noetic quality (sense of gaining authoritative knowledge), transiency (brief duration), and passivity (sense that something is happening to rather than being controlled by the experiencer).[6]

These characteristics appear across cultures and historical periods, suggesting fundamental features of human consciousness rather than culturally specific constructions. However, the content and interpretation of mystical experiences vary dramatically across traditions. A Christian mystic might encounter Jesus or Mary, while a Hindu practitioner experiences Krishna or Shiva. Buddhist meditators often report encounters with bodhisattvas or insights into emptiness, while shamanic practitioners meet animal spirits or traverse otherworldly landscapes.

Stanislav Grof's extensive research with psychedelic therapy revealed both universal and culturally specific elements in visionary experiences. Common motifs include journey narratives, encounters with entities, symbolic death and rebirth, and geometric or mandala-like visions. However, the specific entities encountered, symbolic content, and interpretive frameworks vary based on personal history, cultural background, and religious training.[7]

Anthropological research supports this pattern of universal structure with culturally specific content. Erika Bourguignon's cross-cultural study of altered states found that while the capacity for mystical experience appears universal, its interpretation and integration into social life varies dramatically across cultures.[8] Some societies view such experiences as pathological, others as divine communication, and still others as natural psychological phenomena requiring no special explanation.

The concept of ego death—the temporary dissolution of self-awareness—appears across traditions under different names: fana in Sufism, samadhi in Hinduism, mystical union in Christianity, and bardo states in Tibetan Buddhism. Despite different theoretical frameworks, these traditions recognize similar phenomenological territory involving the temporary suspension of ordinary self-awareness and its replacement by expanded or cosmic consciousness.

Contemporary research suggests that cultural frameworks don't just interpret mystical experiences but actually shape their occurrence and content. Set and setting—the psychological state and environmental context surrounding a mystical experience—significantly influence both its phenomenology and aftereffects. This indicates that subjective realities emerge from the dynamic interaction between neurobiology, psychology, and culture rather than being purely determined by any single factor.

Psychopharmacology and the Mystical

The renaissance of psychedelic research has provided unprecedented opportunities to study mystical experiences under controlled laboratory conditions. This research reveals both the reliability of chemically induced mystical states and their profound similarities to spontaneous religious experiences throughout history.

Psilocybin research conducted by Roland Griffiths and colleagues represents the gold standard for studying mystical experiences scientifically. In carefully controlled studies with healthy volunteers, approximately 60-70% of participants receiving high doses of psilocybin report complete mystical experiences, while control groups receiving placebos rarely report such phenomena.[9] The experiences often involve ego dissolution, unity with the universe, encounters with sacred beings, and insights about the nature of reality.

These chemically induced experiences aren't diminished versions of "real" mystical states but appear phenomenologically identical to spontaneous mystical experiences reported throughout religious history. Participants often describe them as more real than ordinary reality, with lasting positive effects on well-being, spiritual attitudes, and life meaning. Long-term follow-up studies show that many participants continue to rate their psilocybin experiences among the most meaningful and spiritually significant of their lives years later.

DMT research, conducted by Rick Strassman and others, reveals particularly dramatic visionary states characterized by encounters with apparently autonomous entities in otherworldly environments.[10] These experiences often involve complex interactions with beings described as having their own personalities, agendas, and forms of communication. The consistency of these reports across participants raises intriguing questions about whether such experiences reveal hidden aspects of consciousness or represent sophisticated hallucinations generated by altered brain chemistry.

LSD studies, pioneered by researchers like Stanislav Grof before being curtailed in the 1970s, documented profound consciousness-altering effects that often included mystical elements alongside psychological and perceptual changes. Contemporary research by Robin Carhart-Harris and others has confirmed LSD's ability to produce mystical-type experiences while revealing the neural mechanisms underlying these effects.[11]

The therapeutic applications of psychedelics in treating depression, PTSD, and addiction appear to depend partly on their ability to produce mystical experiences. Studies consistently show that participants who report more complete mystical experiences during psychedelic therapy sessions show greater improvement in psychological symptoms and quality of life measures. This suggests that mystical experiences may have inherent healing properties regardless of their ultimate ontological status.

However, critics caution against conflating chemically induced states with spontaneous religious experiences. Skeptics argue that psychedelics simply hijack normal brain functions to produce convincing but ultimately illusory experiences. The fact that mystical experiences can be triggered by molecules raises questions about their authenticity and spiritual significance for some religious practitioners.

Prominent Thinkers and Research Centers

The scientific study of mystical experience has evolved through contributions from researchers spanning neuroscience, psychology, anthropology, and religious studies. Understanding their different approaches illuminates ongoing debates about how to study subjective realities scientifically.

William James at Harvard University established the foundation for empirical study of religious experience through his systematic documentation of mystical states and their characteristics. His *Varieties of Religious Experience* remains influential for its phenomenological approach and its argument that religious experiences should be evaluated by their fruits rather than their origins.[12]

Andrew Newberg at Thomas Jefferson University pioneered neurotheology through neuroimaging studies of meditation and prayer. His research revealed consistent brain activity patterns during mystical states, though critics argue that correlation doesn't establish causation and that reducing spiritual experience to brain states misses essential aspects of religious life.[13]

Robin Carhart-Harris at the University of California, San Francisco, leads contemporary psychedelic research that combines rigorous neuroscience with careful attention to subjective experience. His work on the default mode network and psychedelic consciousness has advanced understanding of how mystical experiences arise and what they might reveal about normal consciousness.[14]

Roland Griffiths at Johns Hopkins University established the modern paradigm for studying mystical experiences through controlled psychedelic research. His careful methodology and attention to set and setting have produced some of the most robust data on mystical experience while respecting participants' subjective reports.[15]

Stanislav Grof, working initially at Johns Hopkins and later independently, conducted extensive research with LSD and developed theoretical frameworks for understanding non-ordinary states of consciousness. His work bridged clinical psychology, spiritual traditions, and consciousness research, though his theoretical conclusions remain controversial within mainstream academia.[16]

Contemporary research centers include the Johns Hopkins Center for Psychedelic and Consciousness Research, the Imperial College London Centre for Psychedelic Research, and the Multidisciplinary Association for Psychedelic Studies (MAPS). These institutions combine rigorous scientific methodology with respect for subjective experience and cultural diversity.

Religious studies scholars like Huston Smith, Walter Stace, and Ralph Hood have developed frameworks for understanding mystical experience that emphasize both universal features and cultural specificity. Their work provides essential context for interpreting neuroscientific findings within broader human meaning-making systems.

Contemporary Debates and Methodological Challenges

The scientific study of mystical experience faces several fundamental challenges that reflect broader issues in consciousness research and the relationship between subjective and objective knowledge.

The reductionism debate centers on whether scientific explanations of mystical experience diminish their spiritual significance. Reductive materialists argue that understanding the neural basis of mystical states reveals them to be sophisticated illusions generated by altered brain chemistry. Religious traditionalists often resist this interpretation, arguing that spiritual experiences transcend purely physical explanations.

However, many researchers and religious practitioners take middle positions. Some argue that understanding neural mechanisms doesn't invalidate spiritual significance any more than understanding the physics of music invalidates aesthetic experience. Others propose that neuroscience might reveal how the brain accesses spiritual realities rather than how it creates spiritual illusions.

The universalism versus constructivism debate addresses whether mystical experiences reveal universal features of consciousness or are entirely shaped by cultural expectations. Walter Stace and Ralph Hood argued for a universal core of mystical experience that transcends cultural differences, while critics like Steven Katz argue that mystical experiences are so thoroughly shaped by cultural and linguistic frameworks that no culture-independent core exists.[17]

Contemporary research suggests that both positions contain partial truths. Certain phenomenological features appear universal, while content and interpretation vary culturally. This parallels findings in other areas of psychology where universal cognitive mechanisms interact with cultural variation to produce both commonalities and differences across populations.

The authenticity question asks whether chemically induced mystical experiences are equivalent to spontaneous religious experiences. Some religious traditions view any artificially induced state as inauthentic or spiritually meaningless. Others see psychedelics as valid spiritual tools, comparable to meditation, fasting, or other consciousness-altering practices.

Research suggests that phenomenologically, chemically induced and spontaneous mystical experiences are often indistinguishable. However, the cultural meaning and spiritual significance attributed to these experiences may depend on their origin and context. This raises questions about the relationship between neurochemical mechanisms and spiritual authenticity.

Methodological challenges in studying mystical experience include the difficulty of replicating spontaneous experiences in laboratory settings, the subjective nature of the phenomena being studied, and the potential for researcher bias in interpreting results. Unlike other areas of neuroscience, mystical experience research must grapple with questions of meaning, value, and ultimate reality that resist purely objective analysis.

The interpretation of altered brain states during mystical experience remains contentious. Do decreased default mode network activity and increased connectivity represent the brain accessing normally hidden aspects of reality, or do they simply reflect altered information processing that generates convincing illusions? Current neuroscience cannot definitively answer this question, leaving room for both materialist and spiritualist interpretations.

Cross-Disciplinary Connections and Implications

The study of mystical experience connects to numerous other fields, creating opportunities for integrated understanding while raising challenges for traditional disciplinary boundaries.

Anthropology provides essential context for understanding how different cultures integrate mystical experiences into social life. Cross-cultural research reveals both universal patterns and dramatic variations in how societies understand, value, and regulate altered states of consciousness. This work challenges Western assumptions about the pathological nature of unusual experiences while revealing the diversity of human consciousness.

Cognitive psychology contributes understanding of how expectation, attention, and memory influence mystical experiences. Research on suggestion, priming, and false memory formation helps distinguish genuine mystical phenomena from artifacts of psychological processes. However, this research also reveals how all experience, including ordinary consciousness, is actively constructed rather than passively received.

Philosophy of religion grapples with questions about the truth value and spiritual significance of mystical experiences. If such experiences can be induced by chemicals or meditation techniques, what does this say about their ability to reveal ultimate reality? How should religious traditions incorporate scientific findings about mystical states into their theological frameworks?

Neuropharmacology reveals the chemical mechanisms underlying mystical experiences while raising questions about the relationship between matter and mind. The fact that simple molecules can produce profound spiritual experiences challenges dualistic views of consciousness while potentially supporting materialist interpretations of religious phenomena.

Clinical psychology and psychiatry explore the therapeutic applications of mystical experiences while addressing concerns about their potential risks. Research on psychedelic therapy reveals that mystical experiences can have profound healing effects for various mental health conditions, though they can also trigger psychological difficulties in vulnerable individuals.

The study of mystical experience also connects to broader questions in consciousness research about the relationship between brain states and subjective experience, the nature of self-awareness, and the possibility of expanded forms of consciousness. These connections suggest that understanding mystical states might contribute to resolving fundamental puzzles about the nature of mind.

Relevance to Part II: Frameworks for the Extraordinary

Understanding mystical experience provides essential conceptual tools for approaching anomalous phenomena without prejudging their ultimate nature. The frameworks developed for studying religious and spiritual experiences offer sophisticated approaches to subjective realities that transcend simple materialist-spiritualist divisions.

The recognition that profound experiences can emerge from altered brain states while retaining personal and cultural significance provides a model for understanding other unusual experiences. This approach neither dismisses extraordinary claims nor accepts them uncritically but seeks to understand how such experiences arise and what they might reveal about human consciousness and reality.

The relationship between universal phenomenological patterns and cultural interpretation offers tools for evaluating reports of anomalous experiences across different contexts. By understanding how culture shapes the interpretation and memory of extraordinary experiences, we can better distinguish potentially genuine phenomena from cultural artifacts while respecting the significance these experiences hold for those who have them.

The neuroscientific study of mystical states reveals that human consciousness is far more flexible and capable of dramatic alteration than everyday experience suggests. This expanded understanding of consciousness provides conceptual space for considering experiences that transcend ordinary limitations while maintaining scientific rigor in their investigation.

The research on set and setting demonstrates how context profoundly influences extraordinary experiences, providing frameworks for understanding why anomalous phenomena might occur in some circumstances but not others. This understanding can inform both the investigation of such phenomena and the conditions under which they might be studied scientifically.

The recognition that mystical experiences often involve encounters with apparently autonomous entities, expanded awareness, and access to information not available through normal channels provides precedents for considering similar claims in other domains. While this doesn't validate any particular extraordinary claim, it suggests that human consciousness may be capable of experiences that transcend conventional scientific explanation.

Most importantly, the study of mystical experience demonstrates that rigorous scientific investigation can be combined with respect for subjective meaning and cultural significance. This approach provides a model for investigating anomalous phenomena that maintains both scientific rigor and openness to possibilities that challenge current understanding.

The frameworks developed for studying mystical experience prepare us to approach extraordinary claims with appropriate sophistication about the nature of consciousness, the relationship between brain states and subjective experience, and the complex interactions between neurobiology, psychology, and culture. Whether applied to near-death experiences, apparent memories of previous lives, or encounters with seemingly non-physical entities, these frameworks offer tools for investigation that honor both scientific methodology and human meaning-making.

Chapter 6: Systems and Networks— Patterns Behind the Unexplainable

Network Neuroscience, Collective Intelligence, Emergent Group Behavior

In 2008, researchers at Nagoya University in Japan designed an elegant experiment to understand how traffic jams form spontaneously.[1] They placed 22 cars on a circular track with no exits, obstacles, or variations in road conditions. Each driver was instructed to maintain a steady 30 kilometers per hour while staying a safe distance from the car ahead. Despite these simple, uniform conditions, something remarkable happened within minutes: clusters of cars began slowing down and speeding up in waves, creating stop-and-go traffic patterns that propagated around the track like ripples in a pond.

No single driver intended to create a traffic jam. No external factor caused the slowdown. Yet the collective behavior of the system inevitably produced a phenomenon that felt purposeful and structured despite arising from purely local interactions. Each driver was simply responding to the car immediately ahead, but their combined responses created large-scale patterns that seemed to have their own logic and persistence.

This experiment reveals a fundamental principle that operates across scales from cellular networks to social systems: complex, seemingly intelligent patterns can emerge spontaneously from simple rules followed by individual agents. These emergent patterns often exhibit properties that cannot be predicted from understanding individual components alone, challenging our intuitions about causation and suggesting new ways to understand phenomena that appear mysterious or unexplainable.

The traffic jam experiment serves as a metaphor for countless other systems where local interactions generate global patterns. Neurons in the brain coordinate to produce consciousness without any central coordinator. Ant colonies solve complex logistical problems despite individual ants having no understanding of the overall strategy. Financial markets exhibit boom-and-bust cycles that no single trader intends. Social movements spread across populations through networks of personal connections, creating cultural changes that transform entire societies.

Understanding these network dynamics provides essential tools for approaching anomalous phenomena that seem to transcend individual capabilities or conventional causal explanations. If ordinary systems can produce behavior that appears purposeful, coordinated, or even intelligent without central control, then perhaps seemingly extraordinary phenomena might emerge from similar but less understood network processes.

Core Concepts: Networks as Organizing Principles

Network science provides a mathematical framework for understanding how relationships between components give rise to system-level properties. A network consists fundamentally of nodes (individual elements) connected by edges (relationships between elements). This simple structure can represent virtually any complex system: neurons connected by synapses, people connected by social relationships, genes connected by regulatory interactions, or ideas connected by logical associations.

The power of network analysis lies in revealing how structure influences function. Random networks, where connections form without pattern, behave very differently from networks with specific organizational principles. Scale-free networks, discovered by Albert-László Barabási, contain a few highly connected hubs and many nodes with few connections.[2] This structure appears throughout nature, from the internet to protein interaction networks to social systems, suggesting fundamental principles of organization that transcend specific domains.

Small-world networks, characterized by Duncan Watts and Steven Strogatz, combine local clustering with long-range connections, enabling both efficient local processing and rapid global communication.[3] This architecture appears in brain networks, social networks, and many other complex systems, providing an optimal balance between local specialization and global integration.

Network dynamics reveal how information, influence, or resources flow through these structures. Synchronization phenomena demonstrate how individual oscillators can spontaneously coordinate their behavior through network connections. Steven Strogatz's research shows that fireflies flashing in unison, heart cells beating in rhythm, and even power grid fluctuations follow similar mathematical principles of coupled oscillator synchronization.[4]

Phase transitions in networks describe how small changes in connectivity or coupling strength can produce dramatic shifts in system behavior. Below a critical threshold, network components remain uncoordinated. Above the threshold, global coordination emerges suddenly and robustly. These transitions help explain how coherent patterns can appear and disappear in complex systems, from neural consciousness to social movements.

Cascading failures illustrate how local disturbances can propagate through networks to produce system-wide breakdowns. The 2008 financial crisis exemplified this phenomenon, where problems in one sector of the financial network spread rapidly throughout the global economy. Understanding cascade dynamics reveals both the fragility and resilience of networked systems.

Network neuroscience has revolutionized understanding of brain function by revealing how distributed neural networks coordinate to produce cognition, memory, and consciousness. Rather than treating the brain as a collection of specialized regions, this approach examines how information flows between areas and how different network configurations support different cognitive functions.[5]

Collective Intelligence and Emergent Coordination

Collective intelligence represents one of the most striking examples of network-level phenomena that transcend individual capabilities. This concept describes how groups can solve problems, process information, or make decisions more effectively than their individual members could achieve alone.

Biological examples of collective intelligence abound throughout nature. Ant colonies demonstrate sophisticated problem-solving capabilities through simple local rules followed by individual ants. When foraging, ants lay down pheromone trails that attract other ants, creating positive feedback loops that identify optimal paths to food sources. The colony as a whole solves complex optimization problems—finding shortest routes, allocating labor efficiently, regulating population—without any central planning or individual ant understanding the overall strategy.[6]

Bee colonies exhibit similar collective intelligence in nest site selection, where scout bees explore potential locations and communicate their assessments through waggle dances. The colony ultimately converges on optimal choices through a distributed decision-making process that weighs multiple factors and reaches consensus without central authority.

Swarm robotics engineers apply these biological principles to create artificial systems capable of collective intelligence. Simple robots following basic local rules can coordinate to map unknown environments, transport large objects, or perform search and rescue operations. Michael Rubenstein's thousand-robot swarms demonstrate how sophisticated group behaviors can emerge from minimal individual capabilities and local communication.[7]

Human collective intelligence manifests in various forms, from traditional practices like barn raising and communal harvests to modern phenomena like open-source software development, Wikipedia editing, and crowdsourced scientific projects. These examples show how distributed groups can accomplish tasks that exceed the capabilities of any individual participant.

Digital technologies have amplified human collective intelligence capabilities through platforms that connect large numbers of people and aggregate their contributions. Prediction markets harness the wisdom of crowds to forecast future events more accurately than expert predictions. Citizen science projects enable millions of volunteers to contribute to astronomical observations, species identification, and protein folding research.

However, collective intelligence also exhibits pathological forms where group dynamics produce outcomes inferior to individual reasoning. Groupthink, information cascades, and herd behavior can lead to poor decisions and systematic errors. Understanding both the positive and negative aspects of collective intelligence provides insights into when and how groups transcend individual limitations.

Network effects explain how collective intelligence emerges from the interaction of individual contributions. Diversity of perspectives, decentralized decision-making, and mechanisms for aggregating distributed information appear crucial for effective collective intelligence. When these conditions are met, groups can exhibit problem-solving capabilities that seem to exceed the sum of their parts.

Information Flow and Hidden Connections

Networks enable information transmission across space and time through mechanisms that often remain hidden from direct observation. Understanding these information flow patterns reveals how apparently separate events might be connected through complex causal chains or how past events might influence present circumstances through indirect pathways.

Epigenetics represents one of the most significant discoveries in modern biology, revealing how environmental experiences can influence gene expression patterns that persist across generations. Unlike genetic mutations that alter DNA sequences, epigenetic modifications change how genes are expressed without changing the underlying genetic code. These modifications can be influenced by factors such as diet, stress, toxin exposure, and social experiences.[8]

Research on transgenerational epigenetic inheritance has revealed that some environmentally induced epigenetic changes can be transmitted to offspring, potentially explaining how ancestral experiences might influence descendant characteristics. Studies of Holocaust survivors and their children suggest that extreme trauma can produce epigenetic changes that persist across generations, potentially affecting stress responses and mental health.[9]

While the mechanisms and extent of transgenerational epigenetic inheritance remain areas of active research and debate, these findings challenge traditional boundaries between individual and inherited characteristics. They suggest that information about environmental experiences can be encoded in biological systems and transmitted across generations through non-genetic mechanisms.

Social networks enable information transmission across populations through mechanisms that create apparent coincidences or connections between distant events. Social media platforms, professional networks, and informal social connections create pathways for information to spread rapidly across vast distances and through diverse populations.

Weak ties, connections between individuals who are not closely related socially, play crucial roles in information transmission. Mark Granovetter's research revealed that people often learn about job opportunities through acquaintances rather than close friends, demonstrating how weak network connections facilitate access to non-redundant information.[10]

Six degrees of separation experiments, originally conducted by Stanley Milgram and recently replicated using digital technologies, reveal how surprisingly short paths connect seemingly unrelated individuals in large populations.[11] These findings suggest that information, influence, or resources can travel between distant parts of social networks through shorter paths than intuition suggests.

Economic networks create hidden connections between distant markets, industries, and geographical regions. Global supply chains link local events to distant consequences through complex networks of production, distribution, and financial relationships. The 2011 tsunami in Japan, for example, disrupted global automobile and electronics production through supply chain networks that connected Japanese manufacturers to worldwide markets.

Information networks in the digital age create new forms of hidden connection through algorithmic recommendation systems, search engines, and social media platforms. These systems create filter bubbles and echo chambers that shape information exposure patterns, potentially creating apparent synchronicities or shared experiences among people who have no direct contact but are exposed to similar information streams.

Prominent Thinkers and Research Centers

The development of network science and complexity theory has emerged through contributions from researchers across multiple disciplines, creating new interdisciplinary fields that bridge physics, biology, sociology, and neuroscience.

Steven Strogatz at Cornell University has made fundamental contributions to understanding synchronization phenomena and nonlinear dynamics in complex systems. His research spans from mathematical foundations of coupled oscillators to practical applications in biological rhythms, power grid stability, and social coordination. His popular writing has made complex mathematical concepts accessible to broader audiences while maintaining scientific rigor.[12]

Albert-László Barabási at Northeastern University pioneered the study of scale-free networks and network medicine. His discovery of preferential attachment mechanisms explains why many real-world networks develop highly skewed degree distributions with a few highly connected hubs. His work on network medicine explores how disease genes are connected in cellular networks and how understanding these connections might lead to new therapeutic approaches.[13]

Duncan Watts, currently at Microsoft Research after positions at Columbia University and Yahoo! Research, has contributed to understanding small-world networks, information cascades, and social influence. His research combines rigorous mathematical modeling with large-scale empirical studies of social phenomena using digital data sources.

Dirk Helbing at ETH Zurich applies network and complexity approaches to understanding social systems, crowd dynamics, and urban planning. His agent-based models reveal how individual behavioral rules can produce emergent patterns in traffic flow, pedestrian movement, and economic systems. His work bridges theoretical modeling with practical applications in urban design and crisis management.[14]

The Santa Fe Institute serves as the premier interdisciplinary center for complexity science, bringing together researchers from diverse fields to study emergent phenomena in complex systems. Founded in 1984, the institute has fostered collaborations between physicists, biologists, economists, and social scientists that have advanced understanding of complex adaptive systems.

MIT's Center for Collective Intelligence investigates how people and computers can work together to solve problems more effectively than either could alone. The center's research spans from understanding biological collective intelligence to designing new forms of human-computer collaboration for addressing global challenges.

Network neuroscience has emerged as a distinct field through the work of researchers like Olaf Sporns at Indiana University, who has mapped structural and functional connectivity patterns in the human brain. Ed Bullmore at the University of Cambridge has pioneered applications of network theory to understanding psychiatric disorders and brain development.[15]

Contemporary Debates and Limitations

Network science faces several challenges that reflect broader issues in complexity research and interdisciplinary collaboration. These debates shape how network approaches are applied to understanding both conventional and anomalous phenomena.

The buzzword critique argues that network concepts are sometimes applied metaphorically without rigorous mathematical modeling or empirical validation. Critics contend that calling something a "network" doesn't automatically explain its behavior, and that the proliferation of network metaphors can obscure rather than illuminate underlying mechanisms.

Methodological challenges in network analysis include difficulties in defining appropriate network boundaries, measuring connection strengths, and distinguishing correlation from causation in network relationships. Many network studies rely on static snapshots of dynamic systems, potentially missing crucial temporal patterns and causal relationships.

The reductionism debate questions whether network-level properties are genuinely emergent or simply reflect our inability to track complex interactions among components. Reductionist critics argue that apparent emergence disappears when systems are analyzed with sufficient detail and computational power. Complexity theorists respond that emergence represents fundamental features of organized systems that cannot be eliminated through more detailed analysis.

Scale-free network universality has been questioned by recent research suggesting that many networks previously classified as scale-free may follow different statistical distributions. This debate highlights the importance of careful statistical analysis and the dangers of overgeneralizing network models across domains.

Prediction limitations in complex networks reflect fundamental challenges in forecasting system behavior even when network structure is well understood. Small changes in initial conditions, network topology, or environmental factors can produce dramatically different outcomes, limiting the predictive power of network models.

The causation versus correlation challenge appears throughout network science, where observed correlations between network properties and system behaviors don't necessarily establish causal relationships. Experimental validation of network mechanisms remains difficult in many domains, particularly for large-scale social or economic systems.

Cross-Disciplinary Applications and Connections

Network approaches have found applications across virtually every scientific discipline, creating new opportunities for understanding complex phenomena while revealing fundamental principles that transcend traditional academic boundaries.

Neuroscience applications of network theory have revolutionized understanding of brain function and dysfunction. Connectome mapping projects aim to chart all neural connections in the brain, revealing how network architecture supports cognitive functions. Network approaches to psychiatric disorders suggest that conditions like schizophrenia and depression might result from altered connectivity patterns rather than localized brain lesions.

Social network analysis has revealed hidden patterns of influence, information flow, and resource distribution in human societies. Research on social contagion shows how behaviors, opinions, and health outcomes spread through social networks like infectious diseases, following mathematical models originally developed for epidemiology.

Economic networks connect firms, markets, and countries through trade, financial, and supply chain relationships. Network analysis of these connections reveals how local economic shocks can propagate globally and how network structure influences economic stability and growth. The 2008 financial crisis highlighted the importance of understanding systemic risk in financial networks.

Ecological networks describe feeding relationships, mutualistic interactions, and energy flows in ecosystems. Food web analysis reveals how species extinctions can cascade through ecological networks, leading to unexpected consequences for ecosystem stability. Pollination networks show how plant-pollinator relationships create dependencies that affect both agricultural productivity and biodiversity.

Information networks include the internet, scientific citation networks, and knowledge representation systems. The structure of these networks influences how information spreads, how knowledge accumulates, and how innovations emerge. Search engines and recommendation algorithms depend on network analysis to organize and retrieve information effectively.

Technological networks encompass power grids, transportation systems, and communication networks. Understanding the structure and dynamics of these networks is crucial for maintaining infrastructure reliability and designing resilient systems that can withstand failures and attacks.

Emergence and Self-Organization

Self-organization represents one of the most fundamental concepts in complexity science, describing how ordered patterns arise spontaneously from the interactions of system components without external control or design. Understanding self-organization provides insights into how complex behaviors can emerge from simple rules and how apparent purpose or intelligence can arise without conscious intention.

Biological self-organization appears at every scale from molecular to ecosystem levels. Protein folding demonstrates how linear amino acid sequences spontaneously organize into functional three-dimensional structures guided by physical and chemical forces. Cellular self-organization produces complex intracellular structures and coordinates cellular functions without centralized control.

Developmental biology reveals how multicellular organisms self-organize from single fertilized eggs through processes involving gene regulatory networks, cellular signaling, and physical forces. Pattern formation in developing embryos follows mathematical principles of reaction-diffusion systems, demonstrating how genetic information interacts with self-organizing processes to create biological forms.

Social self-organization explains how human societies develop institutions, norms, and cultural practices without central planning. Language evolution, market formation, and the emergence of social conventions demonstrate how collective behaviors can arise from individual interactions following simple rules or incentives.

Criticality and phase transitions characterize many self-organizing systems, where small changes in parameters can produce dramatic shifts in system behavior. These transitions often occur at critical points where the system exhibits maximum responsiveness to perturbations, suggesting that self-organization tends to produce systems poised at the edge of order and chaos.

The relationship between self-organization and evolution remains an active area of research. Some scientists argue that self-organizing processes constrain and direct evolutionary change, while others contend that evolution shapes self-organizing mechanisms. This debate has implications for understanding the relationship between physical and biological laws in shaping natural phenomena.

Relevance to Part II: Networks as Investigative Tools

Network and systems approaches provide powerful conceptual frameworks for investigating anomalous phenomena without prejudging their ultimate explanations. These tools reveal how apparently mysterious events might emerge from complex but natural processes while remaining open to phenomena that might transcend conventional scientific understanding.

Large-scale coordination and synchronization phenomena demonstrate that groups can exhibit behaviors that transcend individual capabilities through network mechanisms that operate below the threshold of conscious awareness. Understanding these processes provides frameworks for investigating reports of collective experiences, shared intuitions, or coordinated behaviors that seem to exceed normal communication and planning abilities.

Information flow through hidden network pathways suggests mechanisms by which past events might influence present circumstances or how distant events might become correlated through complex causal chains. Epigenetic inheritance, social network effects, and information cascades provide examples of how connections between events might exist despite apparent spatial or temporal separation.

Network dynamics reveal how small local changes can produce large-scale effects and how apparent coincidences might result from underlying network structures rather than random chance. These insights provide tools for distinguishing genuine anomalies from statistical artifacts while revealing how extraordinary-seeming events might emerge from ordinary network processes.

Self-organization and emergence demonstrate that complex, apparently purposeful behaviors can arise without conscious intention or supernatural intervention. This understanding provides frameworks for investigating phenomena that seem to exhibit intelligence, purpose, or coordination beyond what conventional mechanisms might explain, while maintaining scientific rigor in their investigation.

Brain network dynamics during altered states of consciousness suggest mechanisms by which unusual experiences might arise from atypical patterns of neural connectivity and information flow. Understanding how network reconfiguration can produce profound changes in perception, cognition, and subjective experience provides tools for investigating reports of expanded awareness or anomalous perception.

The systems approach emphasizes that phenomena must be understood within their broader contexts rather than as isolated events. This perspective encourages investigation of the environmental, social, and cultural networks within which anomalous experiences occur, potentially revealing previously unrecognized factors that contribute to their occurrence or interpretation.

Most importantly, network and systems approaches provide sophisticated tools for pattern detection and analysis that can reveal hidden structures in complex data. These methods enable rigorous investigation of phenomena that might otherwise remain invisible to conventional analytical approaches while maintaining appropriate statistical standards for distinguishing genuine patterns from random fluctuations.

Network science reminds us that the world is far more interconnected than everyday experience suggests, with hidden relationships that can produce surprising correlations and influences across apparent spatial and temporal barriers. Understanding these connections provides essential background for approaching anomalous phenomena with appropriate sophistication about the complex systems within which all human experience occurs.

Interlude: "Ok, You Obviously Wonder About Ghosts..."

Let's address the ectoplasmic elephant in the room.

If you picked up this book hoping for a deep dive into haunted houses, spectral figures in Victorian garb, or ghost-hunting gadgets that light up ominously in the dark, you might be wondering why there's no dedicated chapter on ghosts and apparitions. Perhaps you've been waiting for us to tackle the shadowy figures glimpsed in peripheral vision, the disembodied voices calling names in empty houses, or the cold spots that raise goosebumps in supposedly haunted locations.

The reason for this omission is not that ghost experiences are uninteresting or unworthy of attention—quite the opposite. They represent one of the most persistent, cross-cultural motifs in human history, appearing in ancient Chinese ancestor veneration, Roman household spirits, Mesoamerican day of the dead traditions, and African ancestral beliefs, as well as contemporary Western ghost stories.[1] The universality of these experiences across cultures and throughout history suggests they touch something fundamental about human psychology and perhaps human experience itself.

But when we apply the same rigorous criteria we've used for selecting the phenomena examined in Part II, ghost experiences fall short of the threshold required for sustained, systematic investigation within our framework.

Why Ghosts Aren't Here (and Why They Still Matter)

The exclusion of ghosts from our main investigation reflects methodological considerations rather than a judgment about the reality or significance of the experiences themselves. Our selection process requires phenomena to meet specific standards for empirical reproducibility and interdisciplinary scientific support that ghost experiences, as currently documented, cannot satisfy.

The scientific research base for ghost phenomena remains insufficient for the kind of cross-disciplinary analysis that anchors each Part II chapter. While parapsychology journals have published studies of apparitional experiences, there is no substantial, ongoing research program in mainstream academic institutions that applies rigorous, peer-reviewed methodologies to ghost phenomena. The scattered research that does exist often suffers from methodological weaknesses, lacks replication by independent research groups, and fails to meet the standards of evidence expected in other scientific domains.

Heavy cultural shaping presents another significant challenge for systematic study. Ghost narratives are deeply influenced by local folklore, religious beliefs, historical contexts, and media depictions, making it extremely difficult to isolate common phenomenological features from cultural interpretations. A Victorian ghost story differs dramatically from a Japanese yurei account or an indigenous spirit encounter, not just in details but in fundamental assumptions about the nature of death, consciousness, and the relationship between the living and the dead.

The evidence base remains dominated by anecdotal reports and personal testimony, which, while valuable for understanding subjective experience, are vulnerable to well-documented psychological processes including memory reconstruction, social contagion effects, and cultural priming. Unlike near-death experiences, which can sometimes be studied in controlled medical settings, or childhood past-life memories, which leave documentary trails that can be investigated, ghost encounters typically occur under circumstances that resist systematic documentation and verification.

This methodological challenge doesn't invalidate the profound impact these experiences have on those who report them. Ghost encounters can be deeply meaningful, emotionally transformative, and psychologically significant regardless of their ultimate ontological status. The experiences are unquestionably real to those who have them, and they serve important cultural and psychological functions across societies.

What Science Can Say

Even without dedicating a full chapter to ghost phenomena, multiple scientific disciplines have developed relevant frameworks for understanding experiences that people interpret as ghostly encounters. These approaches don't necessarily explain away such experiences but provide naturalistic contexts for understanding how they might arise and why they take the forms they do.

Sleep science has identified several conditions that can produce vivid, seemingly real encounters with presences or figures. Sleep paralysis, occurring during the transition between sleep and waking, can generate intense hallucinations of shadowy figures, feelings of presence, and sensations of being watched or touched while the individual remains conscious but temporarily unable to move.[2] These experiences often occur in familiar environments like bedrooms, adding to their convincing quality.

Hypnagogic and hypnopompic hallucinations—occurring at the edges of sleep—can produce detailed visual, auditory, and tactile experiences that seem entirely real. The brain's predictive processing systems, discussed in Chapter 1, may generate these experiences when normal sensory input is reduced but consciousness remains partially active.

Bereavement psychology has documented that encounters with deceased loved ones are remarkably common during the grieving process. These experiences, reported across cultures, sometimes include visual appearances, auditory communications, tactile sensations, or simply strong feelings of presence.[3] While intensely meaningful to those who experience them, they appear to represent normal psychological responses to loss rather than supernatural phenomena.

Environmental psychology has identified various physical factors that can contribute to feelings of unease, presence, or supernatural dread. Infrasound—sound waves below the range of human hearing—can produce feelings of anxiety, unease, and the sense of being watched. Electromagnetic field variations, temperature fluctuations, air currents, and even mold exposure have all been proposed as potential triggers for experiences interpreted as ghostly.[4]

Cognitive psychology offers insights into how expectation, suggestion, and cultural priming can influence perception and memory. When people enter environments with reputations for being haunted, their heightened attention and expectation can lead them to interpret ambiguous sensory input as supernatural phenomena. Memory reconstruction processes can then elaborate and solidify these initial interpretations into convincing narratives of ghost encounters.

Neurobiology research on temporal lobe activity has revealed that electrical stimulation of certain brain regions can produce feelings of presence, out-of-body experiences, and encounters with apparently external entities. While this doesn't explain all ghost experiences, it suggests neurological mechanisms that might contribute to some reports.

Cultural anthropology provides frameworks for understanding how different societies interpret and integrate experiences of apparent contact with the dead. These studies reveal both universal psychological needs served by ghost beliefs and dramatic cultural variations in how such experiences are understood, valued, and integrated into social life.

Connections to Investigated Phenomena

While ghost experiences don't merit their own chapter, they overlap significantly with phenomena we do examine in detail. Understanding these connections helps explain why we focus on certain experiences while setting others aside.

The predictive brain framework from Chapter 1 provides tools for understanding how expectation and context might generate convincing perceptual experiences in the absence of external stimuli. When the brain's predictive models run without adequate sensory correction—in darkened environments, during stress, or in locations primed with supernatural expectations—they might generate experiences that seem externally caused.

Systems and network dynamics from Chapter 6 offer perspectives on how ghost beliefs and experiences spread through communities via social contagion and cultural transmission. The network effects that can create meaningful coincidences and shared experiences might also contribute to the clustering of ghost reports in particular locations or time periods.

Mystical and altered states discussed in Chapter 5 share phenomenological features with some ghost encounters, including feelings of presence, ego boundary dissolution, and contact with apparently autonomous entities. The cultural framing mechanisms that shape mystical experience interpretation also influence how people understand and remember potential ghost encounters.

Near-death experiences, which we examine in Chapter 7, sometimes include encounters with deceased relatives or spiritual beings that share features with ghost encounters. However, NDEs occur in documented medical contexts that allow for systematic study, while ghost encounters typically lack such controlled circumstances.

Research Standards and Selection Criteria

The decision to exclude ghosts from detailed investigation reflects our commitment to consistent methodological standards rather than prejudgment about the ultimate nature of these experiences. Each phenomenon examined in Part II meets three stringent criteria that guide our investigation.

First, substantial ongoing scientific research must exist, conducted by qualified scholars using peer-reviewed methodologies and published in reputable academic venues. This research need not reach definitive conclusions, but it must demonstrate sustained, serious engagement with the phenomenon using recognized scientific approaches. The goal is not to require proof of extraordinary claims but to ensure sufficient empirical foundation for meaningful analysis.

Second, cross-cultural empirical replication must demonstrate that the phenomenon appears across multiple cultural contexts and historical periods with recognizable consistency. This criterion helps distinguish potentially universal aspects of human experience from culturally specific beliefs or practices. Universal elements suggest phenomena that might reflect fundamental features of consciousness, perception, or reality, while purely cultural elements are better understood through anthropological rather than scientific investigation.

Third, observable patterns must exist that can be documented, compared, and analyzed using systematic methods. These patterns might involve physiological correlates, behavioral consequences, environmental factors, or other measurable characteristics that allow for hypothesis testing and theory development. The requirement isn't for complete understanding but for sufficient structure to support scientific investigation.

Ghost experiences easily meet the second criterion, appearing across virtually all human cultures and historical periods. They partially meet the third criterion, showing some recurring features like feelings of presence, environmental cold spots, and emotional associations with specific locations. However, they currently fail the first criterion in any sustained way, lacking the substantial research infrastructure necessary for the interdisciplinary analysis that defines our approach.

This exclusion is not permanent or absolute. If major research programs devoted to systematic investigation of apparitional phenomena emerge in academic institutions—employing rigorous methodologies, independent replication, and peer review—then ghost experiences might well qualify for the kind of detailed analysis we apply to other phenomena. Science has surprised us before by legitimizing previously dismissed topics, and it might do so again.

The Bigger Picture

Excluding ghosts from our main investigation while acknowledging their cultural significance and psychological reality exemplifies the balanced approach we take throughout this book. We aim to maintain rigorous empirical standards while respecting the complexity and importance of human experience that challenges conventional scientific explanation.

This approach recognizes that science is most effective when it focuses on phenomena that can be studied systematically while remaining open to expanding its domain as new methods and opportunities arise. By applying consistent criteria to all phenomena under consideration, we can explore the edges of current scientific understanding without abandoning the methodological rigor that makes scientific investigation productive.

The exclusion of ghosts also highlights an important distinction between personal meaning and scientific investigation. Experiences can be profoundly significant and transformative for individuals while not meeting the criteria for systematic scientific study. This doesn't diminish their importance or suggest they should be dismissed, but rather acknowledges that different types of human experience require different approaches to understanding.

Our focus on phenomena that can be studied scientifically shouldn't be interpreted as materialist reductionism or skeptical debunking. Rather, it represents a strategic choice to concentrate investigative resources where they can be most productive while remaining open to mystery and wonder. By understanding how science can approach the extraordinary, we better appreciate both its capabilities and its limitations.

The real spirit of inquiry lies in studying what can be measured and compared while maintaining intellectual humility about the boundaries of current knowledge. Ghost experiences remind us that human experience extends far beyond what current science can explain, serving as a humbling reminder of how much mystery remains. Whether these experiences reflect psychological processes, cultural phenomena, or something more extraordinary, they point toward the vast territories of human experience that await future investigation.

As we turn to Part II and its examination of phenomena that do meet our criteria for investigation, we carry forward the lessons learned from considering what to exclude as well as what to include. The boundaries between the scientifically investigable and the perpetually mysterious may shift over time, but maintaining those boundaries thoughtfully allows us to approach the unknown with appropriate tools and realistic expectations.

Part II

Phenomena That Refuse to Go Away

Anomalous experiences reexamined through the lens of modern science. Armed with the theoretical tools from Part I, we now turn to reports that have persisted across cultures and centuries despite scientific skepticism: consciousness during clinical death, children's apparent memories of previous lives, dreams that seem to anticipate future events, telepathic connections between distant minds, encounters with entities during altered states, and extraordinary mental abilities that emerge without explanation. These phenomena resist easy dismissal, yet they also resist easy acceptance. By applying rigorous analysis while remaining open to mystery, we can explore what these experiences might reveal about the hidden depths of consciousness and reality.

Chapter 7: Near-Death Experiences — Consciousness Beyond the Brain?

In 2008, a 57-year-old man suffered cardiac arrest in a hospital ward in Southampton, England. His heart stopped beating for three minutes. Medical staff immediately began CPR and used a defibrillator to restart his heart. During this time, monitors showed no detectable brain activity—a flat EEG that should indicate unconsciousness. Yet when the patient regained consciousness several days later, he astonished the medical team by providing an extraordinarily detailed account of the resuscitation attempt. He accurately described the automated voice from the defibrillator, the specific actions of individual staff members, and even conversations that occurred while he was clinically dead.[1]

This case emerged from the AWARE study (AWAreness during REsuscitation), the largest prospective investigation of consciousness during cardiac arrest, led by critical care physician Sam Parnia at Stony Brook University.[2] While only a small percentage of patients in the study reported any memories from the period of cardiac arrest, those who did often described experiences that challenge our fundamental assumptions about the relationship between brain and consciousness.

Defining the Phenomenon

A near-death experience represents one of the most documented yet controversial phenomena at the intersection of

113

consciousness research, medicine, and human mystery. Psychiatrist Bruce Greyson of the University of Virginia, who has studied NDEs for over four decades, defines them as "profound psychological events with transcendental and mystical elements, typically occurring to individuals close to death or in situations of intense physical or emotional danger."[3]

Greyson's NDE Scale, developed in 1983 and still considered the gold standard for research, identifies core features that appear with remarkable consistency across cases. These include the sensation of leaving one's physical body and observing the scene from above, movement through a tunnel or passage toward an intense light, encounters with deceased relatives or spiritual beings, a comprehensive review of one's life events, radically altered perception of time, and overwhelming emotions— typically profound peace, unconditional love, or unity with the universe.[4]

What makes NDEs particularly intriguing to researchers is not just their internal consistency, but their occurrence during states when brain function should be severely compromised or absent entirely. Cardiac arrest, the most commonly studied trigger, involves the cessation of blood flow to the brain, which typically results in unconsciousness within ten to twenty seconds.[5]

Cross-Cultural Patterns and Variations

The phenomenon transcends cultural and religious boundaries in ways that both support and complicate various explanatory models. Cardiologist Pim van Lommel's landmark study, published in *The Lancet* in 2001, followed 344 cardiac arrest survivors across Dutch hospitals and found that 18% reported NDEs, regardless of their religious background or prior beliefs about afterlife.[6]

Yet cultural context clearly shapes the imagery and interpretation of these experiences. In Satwant Pasricha's studies of NDEs in India, experiencers often report being taken before a divine figure who consults a book of records, only to discover that the wrong person has been brought and they must return to life.[7] Indigenous North American accounts frequently involve traversing natural landscapes or receiving guidance from animal spirits rather than the tunnel-and-light imagery common in Western reports.[8]

Anthropologist Gregory Shushan, who has conducted the most comprehensive cross-cultural analysis of NDE-type experiences, argues that while the surface imagery varies dramatically, the underlying phenomenological structure remains remarkably stable across cultures and historical periods. This suggests either a universal neurological basis for such experiences or, as some researchers propose, access to genuine transpersonal realms that are culturally interpreted through familiar symbolic frameworks.[9]

The Scientific Debate: Three Distinct Positions

As research tools from neuroscience, consciousness studies, and quantum physics have been applied to NDE investigation, three distinct positions have emerged in the scientific community, each with sophisticated arguments and notable limitations.

Physicalist Skeptics: Brain-Based Explanations

The mainstream scientific position holds that NDEs, however profound subjectively, result entirely from brain processes during extreme physiological stress. Susan Blackmore, a psychologist at the University of Plymouth and prominent NDE skeptic, argues that the experiences can be fully explained through known neurological mechanisms operating under crisis conditions.[10]

Blackmore and her colleagues point to several converging lines of evidence. The tunnel vision commonly reported in NDEs closely resembles the visual effects of reduced blood flow to the retina and visual cortex. The life review might result from random firing in the temporal lobe, where autobiographical memories are stored. The profound emotional content could reflect the brain's release of endorphins and other neurochemicals during stress.[11]

Anesthesiologist Gerald Woerlee has provided particularly detailed physiological explanations for veridical perception cases—instances where patients report accurate observations of events during their cardiac arrest. Woerlee argues that even during apparent unconsciousness, patients may retain some sensory awareness through hearing, touch, or brief moments of partial consciousness that create islands of memory formation.[12]

When confronted with cases like the Southampton patient who described events with unusual specificity, physicalist skeptics emphasize that "flat EEG" does not guarantee complete brain inactivity. Deeper brain structures may continue functioning below the threshold of surface electrical detection, potentially maintaining minimal consciousness and sensory processing.[13]

Methodological Critics: Questioning the Evidence

A second group of researchers, while remaining within conventional scientific frameworks, raises pointed questions about the methodology and interpretation of NDE research. These critics don't necessarily deny that something unusual occurs, but they challenge whether current evidence supports extraordinary claims about consciousness surviving brain death.

Psychologist Chris French of Goldsmiths, University of London, exemplifies this position. French acknowledges that some NDE features resist easy explanation but argues that the research methodology often fails to meet rigorous scientific standards. Many studies rely on retrospective reports collected days, weeks, or even years after the experience, introducing significant memory distortion. The verification of veridical perceptions is often incomplete, and alternative explanations are not systematically ruled out.[14]

Neurologist Kevin Nelson of the University of Kentucky has conducted some of the most sophisticated physiological research on NDEs. His work suggests that the experiences may result from the intrusion of REM sleep consciousness into the waking state during times of extreme physiological stress—a phenomenon called REM intrusion that can explain many NDE features including out-of-body experiences, vivid imagery, and altered time perception.[15]

When methodological critics examine cases like the AWARE study, they note that out of 2,060 cardiac arrest patients, only 330 survived and were well enough to interview, and of those, only 140 reported any memories of the period of unconsciousness. Of these, just 9 reported traditional NDE features, and only 2 described any potentially veridical perceptions. This extremely low rate leads critics to question whether the phenomenon represents genuine anomalous consciousness or highly selective reporting of coincidental accurate guesses.[16]

Anomaly Proponents: Evidence for Non-Local Consciousness

A third group of researchers, while maintaining scientific rigor, argues that accumulating evidence points toward consciousness operating independently of normal brain function during some NDEs. These researchers don't necessarily embrace supernatural explanations, but they contend that materialist models cannot adequately account for all NDE features.

Bruce Greyson, Professor Emeritus of Psychiatry and Neurobehavioral Sciences at the University of Virginia, has spent decades documenting NDEs with careful attention to verification procedures. Greyson argues that while many NDE features might be explained through brain physiology, certain cases present evidence that stretches conventional explanations beyond their breaking point. These include instances of enhanced cognitive function during documented periods of brain dysfunction, accurate perceptions of events occurring outside the patient's sensory range, and transformative psychological effects that persist for decades.[17]

Pim van Lommel's prospective study design specifically addressed many methodological criticisms by interviewing patients within days of their cardiac arrest and by including control groups of cardiac patients who did not report NDEs. Van Lommel argues that if NDEs resulted simply from brain dysfunction during cardiac arrest, they should occur in all patients or at least correlate with factors like duration of unconsciousness, medication levels, or degree of brain hypoxia. Instead, NDE occurrence appears largely independent of medical variables, suggesting additional factors beyond mere brain physiology.[18]

Perhaps most compellingly, proponents point to cases of accurate veridical perception that resist conventional explanation. The AWARE study included sophisticated controls, placing visual targets in resuscitation areas that could only be seen from above. While no patient in the initial study reported seeing these targets, researchers argue that the study established important protocols for future investigation and that the rarity of such perceptions doesn't negate their significance when they do occur.[19]

Frameworks from Part I: New Lenses for Ancient Questions

The theoretical frameworks explored in Part I offer fresh perspectives on these debates, though each comes with both insights and limitations when applied to NDE research.

The Predictive Brain: Internal Models Under Stress

The predictive processing model suggests that the brain constantly generates internal simulations of reality, updating them based on incoming sensory data. During the extreme conditions that trigger NDEs—cardiac arrest, severe injury, or drug-induced states—normal sensory input becomes drastically reduced or distorted. Under these conditions, the brain might shift into a fully generative mode, creating immersive experiential worlds from internal models with minimal external constraint.[20]

This framework elegantly explains many NDE features: the tunnel could represent the brain's default visual processing under minimal input conditions, the life review might emerge from the spontaneous activation of autobiographical memory networks, and the profound sense of reality could reflect the brain's inability to distinguish between internally generated and externally derived experience when normal reality-testing mechanisms are offline.

Yet when researchers like van Lommel examine cases of apparently accurate veridical perception during documented periods of brain inactivity, they argue that predictive processing models, while illuminating, cannot fully account for instances where internal models somehow match external events with precision that exceeds chance. Skeptics respond that the apparent precision may be illusory, resulting from selective reporting, memory reconstruction, or brief periods of partial consciousness that escape detection.

Consciousness Models: Fundamental or Emergent?

The hard problem of consciousness, as articulated by David Chalmers, questions how subjective experience emerges from objective brain processes. If consciousness is indeed an emergent property of complex neural activity, then NDEs represent extreme but ultimately brain-bound experiences. However, if consciousness involves fundamental features of reality—as suggested by Integrated Information Theory or panpsychist approaches—then NDEs might represent moments when awareness operates through non-neural mechanisms.[21]

Proponents like van Lommel draw on this framework to argue that consciousness might not be produced by the brain but rather interfaced through it. In this view, severe brain dysfunction doesn't eliminate consciousness but alters its mode of operation, potentially allowing access to information through non-local mechanisms. Critics like Christof Koch counter that while consciousness remains mysterious, there is no credible evidence for awareness operating independently of neural activity, and NDEs likely represent the brain's final coherent experiences before complete shutdown.[22]

The debate reveals a fundamental epistemological challenge: how can we investigate consciousness using tools that assume consciousness is brain-based? As philosopher Thomas Nagel has noted, our current scientific methods may be inherently limited in their ability to address questions about the fundamental nature of consciousness.[23]

Quantum Mechanics and Information

Some researchers have proposed that quantum mechanical processes might enable consciousness to access information through non-classical pathways. Physicist Henry Stapp has suggested that consciousness might involve quantum measurement processes that allow the mind to influence physical events, including potentially accessing information beyond normal sensory channels.[24]

However, most quantum physicists remain deeply skeptical of such applications to consciousness. Sean Carroll argues that quantum effects in warm, wet brain tissue decohere far too quickly to support sustained non-local consciousness. While quantum entanglement can create correlations between distant particles, these correlations cannot transmit specific information and break down rapidly in biological systems.[25]

Roger Penrose and Stuart Hameroff's Orchestrated Objective Reduction theory represents the most developed attempt to connect quantum processes to consciousness, proposing that quantum computations in microtubules within neurons might generate conscious experience. While this theory has gained some experimental support, most neuroscientists consider it highly speculative and unnecessary for explaining observed brain functions.[26]

Artificial Minds and Narrative Generation

The emergence of sophisticated AI systems offers new metaphors for understanding how coherent experiences might emerge from minimal or degraded input. Like advanced language models that can generate compelling narratives from fragmentary prompts, the brain during NDEs might weave detailed, meaningful experiences from minimal sensory data and internal memory fragments.[27]

This perspective suggests that the compelling quality of NDEs—their sense of hyperreality and profound meaning—might reflect the brain's extraordinary capacity for pattern completion and narrative construction rather than access to transpersonal realms. Critics of anomalous explanations find this framework particularly compelling because it accounts for both the subjective richness of NDEs and their occurrence during states of compromised brain function.

Yet proponents note that this framework, while illuminating the brain's capacity for narrative construction, doesn't address cases where the constructed narrative includes accurate information about external events that were allegedly not accessible through normal sensory channels during the experience.

Systems and Networks: Reorganization Under Crisis

Network neuroscience reveals that the brain operates through large-scale, interconnected systems that can reorganize rapidly in response to changing conditions. During the extreme physiological stress that triggers NDEs, these networks might undergo radical reorganization, potentially explaining both the unusual features of the experience and their lasting psychological impact.[28]

Research on psychedelic states, which share phenomenological similarities with NDEs, shows that such experiences involve the breakdown of the brain's "default mode network"—a system associated with self-referential thinking and the maintenance of normal ego boundaries. This breakdown correlates with experiences of ego dissolution, unity consciousness, and altered time perception that are also common in NDEs.[29]

Steven Strogatz's work on synchronization in complex systems suggests that under extreme conditions, large networks can suddenly shift into highly ordered, coherent states. Applied to NDEs, this framework might explain how a brain under severe stress could generate experiences with unusual clarity, integration, and emotional intensity.[30]

However, this systems perspective, like the others, struggles with cases of apparently accurate perception during states of documented brain dysfunction. While network reorganization might explain the subjective qualities of NDEs, it's unclear how such reorganization could enable access to information beyond the normal sensory boundaries of the brain.

The Current State of Evidence: An Honest Assessment

After decades of research, what can be said with confidence about NDEs? The phenomenon itself is well-documented and remarkably consistent across cultures. The experiences have profound, lasting effects on those who report them, often leading to reduced fear of death, increased compassion, and altered life priorities that persist for years or decades.[31]

The physiological correlates of NDEs are becoming clearer. They occur during states of severe brain stress or dysfunction, often when EEG readings show minimal or no detectable activity. Various brain-based mechanisms can account for many specific features of the experience, from tunnel vision to life reviews to out-of-body sensations.

Yet significant questions remain. The cases of apparently accurate veridical perception, while rare, have not been definitively explained by conventional mechanisms. The transformative psychological effects appear disproportionate to what might be expected from brief episodes of brain dysfunction. And the cross-cultural consistency of the experience suggests either universal neurological mechanisms or access to transpersonal dimensions of reality.

Perhaps most importantly, the debate reveals the limitations of current scientific methods when applied to consciousness itself. As philosopher of science Thomas Kuhn noted, scientific paradigms shape not only what questions are asked but what kinds of answers are considered legitimate. The NDE debate occurs at the intersection of neuroscience, consciousness studies, and questions about the fundamental nature of reality—areas where our current paradigms may be insufficient.[32]

Looking Forward: Questions for Further Investigation

Rather than settling the debate, current research has refined the questions that need addressing. If consciousness can operate independently of normal brain function, what mechanisms might enable this? If NDEs are entirely brain-based, how do we

125

account for their apparently anomalous features? How can we design studies that adequately test these competing hypotheses?

The AWARE-2 study, currently underway, represents an attempt to address some methodological limitations of earlier research. By using more sophisticated monitoring of brain activity and improved protocols for testing veridical perception, researchers hope to generate clearer evidence about what occurs during cardiac arrest.[33]

Yet the fundamental epistemological challenges remain. How can we study consciousness using methods that assume consciousness is brain-produced? How do we investigate claims about non-local awareness using instruments designed to detect local physical processes? These questions extend far beyond NDEs to the heart of consciousness research itself.

The debate also raises profound questions about the nature of scientific evidence and the boundaries of legitimate inquiry. What level of proof should be required before considering explanations that challenge current paradigms? How do we balance scientific skepticism with openness to potentially paradigm-shifting evidence?

As we'll see in subsequent chapters, these same epistemological challenges appear throughout the investigation of anomalous phenomena. The tools of science, developed to study the material world, may require fundamental extensions to address questions about consciousness, information, and the nature of reality itself.

The Southampton patient who observed his own resuscitation represents just one data point in this larger investigation. Whether his experience reflects the brain's final creative act or evidence for consciousness beyond the brain remains an open question—one that may ultimately require new frameworks for understanding the relationship between mind and reality.

Cultural Lens: The Interpretation Problem

The cross-cultural variations in NDE imagery highlight a crucial challenge in consciousness research: the problem of cultural interpretation. A Tibetan Buddhist might describe encountering a yidam deity during their experience, while a secular Westerner reports meeting deceased relatives in a tunnel of light. A Native American might describe traveling through natural landscapes guided by animal spirits, while a Hindu might report appearing before Yamaraj, the lord of death, who consults divine records.

Are these fundamentally different experiences, or the same underlying phenomenon filtered through different cultural interpretive frameworks? If the latter, it suggests that while the core neurological or consciousness processes might be universal, their expression is inevitably shaped by cultural conditioning, religious beliefs, and symbolic systems.

This interpretive challenge extends beyond mere imagery to fundamental questions about the nature of the experience itself. Western research tends to focus on individual consciousness and brain function, while many non-Western traditions view such experiences in terms of collective spiritual realities or cosmic consciousness. These different frameworks generate different research questions, methodologies, and interpretive possibilities.

Without robust cross-cultural research that takes these interpretive differences seriously, we risk mistaking culturally specific expressions for universal features of the phenomenon— or conversely, dismissing genuine commonalities because of surface differences in imagery and interpretation.

Chapter 8: Reincarnation — Memory Without the Self

In the early 2000s, a two-year-old boy in Louisiana began having nightmares about being trapped in a burning airplane. James Leininger would wake up screaming, "Airplane crash! Fire! Little man can't get out!" As the episodes continued, more details emerged. He said he had flown from a boat called the *Natoma Bay*, that he had been shot down by the Japanese, and that his plane was a Corsair. He named a fellow pilot, Jack Larsen, who "was there when I got shot down." James also signed his drawings "James 3"—though he was James Leininger Jr., and no one had taught him to add numbers to names.[1]

James's parents, Bruce and Andrea Leininger, were initially skeptical. Bruce, a Christian and former skeptic of reincarnation, began investigating his son's claims to prove them wrong. Instead, he discovered the historical existence of James McCready Huston Jr., a 21-year-old Navy pilot killed in action on March 3, 1945, near Iwo Jima. Huston had indeed flown from the USS *Natoma Bay*, piloted a Corsair fighter, and served alongside a pilot named Jack Larsen. Military records confirmed details that the toddler had somehow known.[2]

This case became one of the most extensively documented in the files of the University of Virginia's Division of Perceptual Studies, which has investigated over 2,500 reports of children who claim memories of previous lives. While the interpretation of such cases remains highly controversial, the systematic nature of the research and the specificity of many reports have made reincarnation one of the most rigorously studied—and hotly debated—anomalous phenomena in contemporary science.

Defining the Research Territory

Reincarnation, as understood in various religious and philosophical traditions, refers to the rebirth of some aspect of personal identity—consciousness, soul, or essential self—in a new physical body after death. Hindu and Buddhist traditions frame this as part of a cosmic cycle of death and rebirth (*samsara*) governed by karmic law, while Platonic philosophy entertained the soul's migration between lives as a means of acquiring knowledge and moral development.[3]

In the scientific context, reincarnation research focuses on a more specific phenomenon: cases where individuals, typically young children, report detailed memories, skills, or physical characteristics that appear to correspond to the life of a deceased person whom they could not have known through normal means. This operational definition allows researchers to investigate specific claims without necessarily endorsing metaphysical frameworks about souls or afterlife.

Ian Stevenson, who founded this field of study at the University of Virginia in the 1960s, established rigorous criteria for case investigation. Strong cases typically include detailed memories provided before any investigation begins, multiple independent witnesses to the child's statements, verification of a deceased person matching the child's descriptions, and no normal means by which the child could have acquired the information.[4]

Patterns Across Cases and Cultures

Jim Tucker, who succeeded Stevenson as director of the reincarnation research program, has identified consistent patterns across thousands of cases spanning six decades of investigation. Children typically begin discussing previous life memories between ages two and five, often with considerable emotional intensity. The memories usually fade as the child reaches school age, generally disappearing by age seven or eight.[5]

The content of these memories often includes specific details about the previous personality's life: names of family members, occupations, locations of homes or workplaces, and circumstances of death. In approximately 70% of cases that can be investigated, the previous personality died a violent or unexpected death, leading some researchers to speculate that traumatic endings may somehow facilitate memory retention across lives.[6]

Many children also display behaviors, preferences, or skills that seem related to their claimed previous life. Some show precocious knowledge of locations they've never visited, speak words from languages they've never been taught, or demonstrate fears or phobias apparently connected to their reported manner of death in the previous life.

Perhaps most intriguingly, about 35% of cases involve birthmarks or birth defects that correspond to wounds or markings on the deceased person. Stevenson documented hundreds of such cases, often obtaining autopsy reports or medical records that confirmed the correspondence between the child's physical features and the previous personality's injuries.[7]

Cross-Cultural Manifestations

While reincarnation cases occur worldwide, their frequency and characteristics vary significantly across cultural contexts. In societies where reincarnation is an accepted belief—such as parts of India, Myanmar, Turkey, and among certain Native American tribes—cases are reported more frequently and are often investigated by family members or community elders.

In India, where Stevenson conducted extensive research, children sometimes report being members of higher or lower castes in their previous lives, details that families can verify through local records. Satwant Pasricha's research documented how Indian cases often involve children claiming to have been merchants or householders, with memories focused on family relationships and business matters.[8]

Among the Druses of Lebanon, a close-knit religious community that strongly believes in reincarnation, children's statements about previous lives are taken seriously and systematically investigated by family members. The community's detailed genealogical records and tight social networks often allow for rapid verification of claims.[9]

In Western countries, where reincarnation is less culturally accepted, cases are reported less frequently but often involve more detailed investigation due to the families' initial skepticism. These cases sometimes attract more rigorous documentation precisely because the families seek to disprove rather than confirm their children's claims.

Anthropologist Antonia Mills has documented reincarnation beliefs and reported cases among Pacific Northwest Native American tribes, where the phenomenon is understood within frameworks of ancestral return rather than individual soul transmigration. In these cultures, children may be recognized as returned ancestors based on birthmarks, behaviors, or statements that connect them to deceased tribal members.[10]

The Scientific Debate: Three Distinct Approaches

The accumulated evidence from reincarnation research has generated three distinct positions within the scientific community, each bringing different methodological assumptions and explanatory frameworks to bear on the phenomenon.

Physicalist Skeptics: Cultural Construction and Memory Artifacts

The mainstream skeptical position, exemplified by philosophers like Paul Edwards and psychologists like Nicholas Spanos, argues that reincarnation cases can be fully explained through known psychological and cultural mechanisms without invoking survival of consciousness beyond death.

Edwards, in his comprehensive critique published in *The Philosophical Quarterly*, argues that reincarnation reports result from a combination of cultural suggestion, parental reinforcement, cryptomnesia (unconscious plagiarism of forgotten information), and the human tendency toward pattern-seeking in random events. He points out that cases occur predominantly in cultures where reincarnation is believed, suggesting cultural priming rather than genuine memory.[11]

Psychologist Susan Blackmore, applying principles from memory research, argues that children's "past life memories" represent confabulations—false memories constructed from fragments of overheard conversations, television programs, or books, then reinforced through family attention and retelling. The predictive brain, she suggests, fills in gaps in these fragmentary inputs to create coherent narratives that feel like genuine memories.[12]

Nicholas Spanos and his colleagues conducted laboratory studies showing how easily false memories of previous lives can be created through hypnotic suggestion, even in skeptical subjects. They argue that similar processes of inadvertent suggestion may operate in family contexts where children's unusual statements are interpreted as evidence of past lives.[13]

When confronted with cases involving verified obscure details, physicalist skeptics emphasize the possibilities of inadvertent information transmission. Children may overhear adult conversations, encounter information through media, or pick up details from community gossip that they later reproduce as "memories." They also point to confirmation bias in investigation—the tendency to notice and remember hits while forgetting misses.

Methodological Critics: Questioning the Evidence Base

A second group of researchers accepts that something unusual may be occurring in some reincarnation cases but raises substantial questions about research methodology and the interpretation of evidence. These critics don't necessarily reject the possibility of survival but argue that current evidence fails to meet rigorous scientific standards.

Psychologist Christopher French of Goldsmiths, University of London, acknowledges that some reincarnation cases present intriguing features but argues that the research methodology contains systematic flaws that undermine its conclusions. The investigation typically begins only after families have already begun interpreting the child's statements as evidence of a previous life, creating opportunities for memory contamination and investigative bias.[14]

French and his colleagues point out that investigators rarely have access to the child's original statements, relying instead on family recollections that may have been unconsciously modified over time. The verification process often involves leading questions and selective attention to confirming details while ignoring discrepancies.

Cognitive psychologist Richard Wiseman raises additional concerns about the statistical evaluation of cases. Even apparently specific details may have higher base rates of occurrence than investigators assume, making seemingly impressive matches less remarkable than they appear. The lack of systematic control studies—comparing verified reincarnation cases with carefully documented false positives—makes it difficult to assess the true significance of apparent matches.[15]

Psychiatrist Harold Lief, while remaining open to anomalous explanations, argues that the case study methodology, however careful, cannot adequately control for all normal sources of information transfer. He notes that the long intervals between a child's initial statements and their formal investigation provide multiple opportunities for memory distortion and suggestion to operate.[16]

When methodological critics examine exemplary cases like James Leininger, they don't dismiss the evidence but call for more rigorous protocols. What would constitute adequate controls? How could investigators rule out all normal sources of information transmission? How might unconscious bias affect both the collection and interpretation of evidence?

Anomaly Proponents: Evidence for Survival

A third group of researchers, while maintaining scientific rigor, argues that accumulating evidence from the strongest reincarnation cases points toward genuine survival of some aspect of personal consciousness beyond bodily death. These researchers don't necessarily embrace religious or metaphysical interpretations but contend that materialist explanations cannot adequately account for all features of the best-documented cases.

Jim Tucker, current director of the University of Virginia's reincarnation research, argues that while many cases may indeed be explained through normal psychological and cultural mechanisms, a significant subset presents features that resist conventional explanation. Tucker has developed statistical methods for assessing the strength of case evidence, considering factors like the specificity of statements, the independence of verification, and the absence of normal information sources.[17]

Tucker points to cases where children provide highly specific, accurate information about deceased persons who lived in distant locations, died before the child's birth, and had no connection to the child's family or community. In some instances, children have led researchers to previously unknown burial sites or provided details confirmed only through obscure historical records.

Ian Stevenson, the field's founder, documented hundreds of cases involving birthmarks or birth defects that corresponded with remarkable precision to wounds on deceased persons. In his most compelling cases, Stevenson obtained autopsy reports, medical records, or postmortem photographs showing exact correspondence between children's physical features and previous personalities' injuries. The odds of such correspondences occurring by chance, Stevenson argued, become astronomically small when multiple features correlate within single cases.[18]

Emily Williams Kelly, also at the University of Virginia, has studied cases involving children who demonstrate knowledge or skills apparently carried over from previous lives. Some children display age-inappropriate knowledge of technical subjects, speak foreign languages they've never been taught, or demonstrate artistic abilities that seem to correlate with their claimed previous life occupations.[19]

When anomaly proponents address skeptical criticisms, they acknowledge that weak cases probably do result from cultural suggestion and false memory. However, they argue that the existence of poor cases doesn't invalidate strong ones. They also point out that the cultural distribution of cases may reflect differences in reporting and recognition rather than differences in occurrence—Western families may be more likely to suppress or ignore children's statements about previous lives.

Proponents also emphasize that their research methodology has evolved to address many skeptical concerns. Modern investigations increasingly involve independent verification by multiple researchers, careful documentation of original statements before investigation begins, and systematic attempts to rule out normal sources of information.

Frameworks from Part I: New Perspectives on Ancient Questions

The theoretical frameworks developed in Part I offer fresh approaches to understanding how reincarnation cases might arise, though each comes with both explanatory power and significant limitations when applied to the accumulated evidence.

The Predictive Brain: Narrative Construction and False Memory

The predictive processing model provides a compelling framework for understanding how children might construct detailed "past life" narratives from fragmentary inputs. If the brain constantly generates internal models of reality based on limited sensory data, children exposed to even minimal information about deceased persons might elaborate these fragments into coherent, memory-like experiences.

Memory researchers like Elizabeth Loftus have demonstrated how easily false memories can be implanted through suggestion, and how these artificial memories often feel as vivid and compelling as genuine ones. Applied to reincarnation cases, this framework suggests that children might unconsciously absorb information from overheard conversations, community stories, or media exposure, then process these inputs through predictive mechanisms that generate seemingly detailed memories.[20]

The model elegantly explains why past life memories often fade as children mature—developing cognitive abilities and expanding real-world experience may override earlier fantasy constructions. It also accounts for the cultural distribution of cases, as societies that expect and interpret unusual childhood statements as evidence of reincarnation would provide more opportunities for such narrative construction.

However, proponents argue that predictive processing models, while illuminating many cases, struggle to explain instances where children provide verifiable information that appears to have no normal source. Tucker and his colleagues have documented cases where children's statements were recorded before any investigation began, where families actively sought to disprove the claims, and where the information provided proved accurate in details that local community knowledge couldn't explain.

Critics respond that even apparently impossible information transfer may result from sources investigators failed to identify. The complexity of information flow in modern societies makes it nearly impossible to rule out all potential channels through which children might acquire knowledge about deceased persons.

Consciousness Models: The Problem of Personal Identity

Theories about the fundamental nature of consciousness raise profound questions about what, if anything, might survive bodily death and be reborn in another body. If consciousness emerges entirely from neural activity, as most neuroscientists believe, then genuine reincarnation becomes essentially impossible—there would be no mechanism for preserving personal memories or identity beyond brain death.

However, if consciousness involves fundamental features of reality that operate partially independently of neural substrates—as suggested by some interpretations of Integrated Information Theory or panpsychist approaches—then reincarnation becomes conceivable, though still requiring explanation of how information transfer might occur.

David Chalmers, while not endorsing reincarnation, has noted that the hard problem of consciousness—explaining how subjective experience arises from objective brain processes—leaves open possibilities that purely materialist approaches might miss. If consciousness involves non-physical properties, those properties might theoretically persist beyond individual brain death.[21]

Philosopher Robert Almeder argues that reincarnation research presents an empirical challenge to materialist theories of mind. If the evidence from strong cases is genuine, it suggests that personal identity involves more than patterns of neural connectivity—something that can retain specific memories and characteristics across different physical embodiments.[22]

Yet even if consciousness were somehow fundamental to reality, critics point out that reincarnation would still require explaining how specific memories, personality traits, and even physical characteristics could be preserved and transmitted from one individual to another. The lack of any known mechanism for such information transfer remains a major obstacle for survival-based theories.

Neuroscientist Christof Koch, while acknowledging consciousness as mysterious, argues that everything science has learned about memory and personal identity points to their dependence on specific brain structures and neural patterns. Without those physical substrates, he contends, there would be nothing to preserve or transmit to a new body.[23]

Quantum Mechanics and Information Transfer

Some researchers have proposed quantum mechanical processes as potential mechanisms for reincarnation, suggesting that consciousness might involve quantum information that could theoretically survive bodily death and influence the development of new organisms.

Physicist Henry Stapp has speculated that consciousness might involve quantum measurement processes that create non-local connections between minds across space and potentially time. In principle, such connections could allow information from deceased persons to influence the neural development or psychological characteristics of children born later.[24]

Stuart Hameroff and Roger Penrose's Orchestrated Objective Reduction theory proposes that consciousness emerges from quantum processes in neural microtubules. While their theory primarily addresses consciousness in living brains, Hameroff has suggested that quantum information might persist beyond individual death in ways that could theoretically influence other conscious systems.[25]

However, most quantum physicists remain deeply skeptical of such applications. Sean Carroll argues that quantum effects in biological systems decohere far too rapidly to support persistent information storage or transmission across death and rebirth. While quantum entanglement can create correlations between distant particles, these correlations cannot transmit specific information like memories or personality traits.[26]

Max Tegmark's calculations suggest that quantum coherence in warm, wet brain tissue lasts for only about 10^{-13} seconds—far too brief to support sustained quantum processes in consciousness, let alone the preservation of information beyond death. The notion that quantum effects could preserve detailed memories across years or decades of bodily decomposition and somehow influence the development of new brains appears to violate basic principles of quantum mechanics.[27]

Even researchers sympathetic to quantum approaches to consciousness, like physicist Freeman Dyson, have noted that quantum mechanics provides no known mechanism for storing or transmitting the complex, specific information that would be required for genuine reincarnation.

Artificial Intelligence and Pattern Matching

The development of sophisticated AI systems offers new metaphors for understanding how children might generate convincing "past life" narratives from minimal input. Large language models can produce remarkably coherent stories based on fragmentary prompts, suggesting how human brains might construct detailed memories from limited information.

Cognitive scientist Douglas Hofstadter has noted that human memory is fundamentally reconstructive rather than reproductive—we don't replay stored recordings but reconstruct experiences from cues and patterns. This reconstructive process could theoretically generate detailed "memories" of events that never occurred in the rememberer's actual life.[28]

Applied to reincarnation cases, this framework suggests that children might unconsciously detect patterns in available information—overheard names, local historical knowledge, community stories—and use these patterns to generate coherent narratives that feel like genuine memories. The brain's extraordinary pattern-completion abilities could fill in gaps to create compelling, detailed stories.

Yet proponents argue that this framework, while explaining the narrative coherence of past life memories, doesn't address cases where children provide specific, verifiable information that appears to have no available source. Unlike AI systems that work with vast databases of information, young children would presumably have access only to limited local knowledge.

Critics respond that children's information environment may be richer than researchers assume. In closely connected communities, historical knowledge may circulate through informal channels that investigators fail to detect. The apparent specificity of some statements may also reflect post-hoc selection—investigators may focus on accurate details while overlooking numerous inaccurate statements.

Systems Theory and Morphic Fields

Network science and systems theory suggest ways that information might be preserved and transmitted through non-obvious channels. Biologist Rupert Sheldrake's controversial hypothesis of morphic resonance proposes that systems can inherit patterns or information from similar past systems through non-physical fields.

Applied to reincarnation, this framework would suggest that individual consciousness might create patterns in morphic fields that could later influence the development of similar conscious systems. Children sharing certain characteristics with deceased persons might become resonant with those morphic patterns, manifesting as apparent memories or behavioral traits.[29]

However, mainstream biologists and physicists have largely rejected morphic field theories as lacking empirical support and contradicting well-established principles of biology and physics. The proposed fields have never been detected, and controlled experiments designed to test morphic resonance have generally failed to produce significant results.

Epigenetics offers a more conventional biological mechanism that might explain some reincarnation-like phenomena. Traumatic experiences can create epigenetic changes that are sometimes transmitted to offspring, potentially influencing behavior, fears, or predispositions in ways that might be misinterpreted as past life memories.

Rachel Yehuda's research on Holocaust survivors and their descendants has shown that trauma can create heritable epigenetic modifications that affect stress responses in children and grandchildren. Such effects might theoretically explain some behavioral or emotional characteristics that children seem to share with deceased persons, particularly in cases involving violent deaths.[30]

Yet epigenetics, while explaining possible predispositions or tendencies, cannot account for specific, detailed memories of particular events, places, or people. The mechanism appears too general to explain the precise correspondences that characterize the strongest reincarnation cases.

Current State of Evidence: An Honest Assessment

After six decades of systematic investigation, what can be concluded about reincarnation research? The phenomenon of children reporting past life memories is well-documented and cross-culturally consistent. Many cases involve details that can be historically verified, and some cases present correspondences—between statements and facts, or between physical features and documented injuries—that exceed what would be expected by chance.

The quality of investigation has improved significantly since Stevenson's early work. Modern researchers use more rigorous documentation procedures, seek independent verification of claims, and attempt to rule out normal sources of information transfer. Video recordings now preserve children's original statements, reducing reliance on family recollections.

However, significant methodological challenges remain. The retrospective nature of most investigations—beginning only after families have interpreted children's statements as evidence of past lives—creates unavoidable opportunities for memory contamination and investigative bias. The lack of prospective studies, where researchers might follow children from birth to identify and document unusual statements before any interpretation occurs, limits the strength of evidence.

The statistical evaluation of cases also presents challenges. Even apparently specific details may have higher base rates than investigators recognize, and the number of potential matches among deceased persons in any region may be larger than initially apparent. Without systematic control studies comparing verified cases with false positives, it remains difficult to assess the true significance of apparent matches.

Cultural factors clearly influence both the reporting and interpretation of cases. The overwhelming concentration of reports in reincarnation-believing societies suggests either that belief affects the occurrence of the phenomenon or that it affects recognition and reporting. Both possibilities have implications for how the evidence should be evaluated.

Perhaps most importantly, no clear mechanism has been identified for how specific memories or personal characteristics could be preserved beyond death and transmitted to new individuals. While various theoretical possibilities have been proposed—from quantum information to morphic fields—none have gained empirical support sufficient to provide a convincing explanation for the alleged phenomenon.

The Interpretive Challenge

The reincarnation debate ultimately confronts fundamental questions about the nature of evidence, the burden of proof, and the boundaries of scientific investigation. How should unusual, well-documented cases be evaluated when they seem to challenge basic assumptions about consciousness and personal identity?

Skeptics argue that extraordinary claims require extraordinary evidence, and that normal explanations should always be preferred over anomalous ones unless absolutely ruled out. They point to the long history of apparently paranormal phenomena that were later explained through conventional means, suggesting that reincarnation cases will eventually yield to similar analysis.

Proponents respond that they are providing extraordinary evidence in the form of detailed case studies with multiple independent verifications. They argue that the consistent dismissal of evidence based on theoretical impossibility rather than empirical evaluation represents a form of scientific prejudice that impedes genuine inquiry.

Methodological critics occupy a middle ground, accepting that something interesting may be occurring while demanding higher standards of evidence and better controls. They ask what kind of evidence would actually be convincing and whether current research methods can provide it.

The debate also raises questions about the relationship between scientific and cultural ways of knowing. In societies where reincarnation is an accepted part of worldview, children's statements are interpreted within frameworks that Western science considers unfounded. Yet these cultural frameworks may preserve important knowledge about human consciousness that purely materialist approaches miss.

How should researchers balance scientific rigor with cultural sensitivity? How can investigation proceed without imposing

Western assumptions about the nature of consciousness and personal identity? These questions extend beyond reincarnation research to broader issues in consciousness studies and cross-cultural psychology.

Future Directions: Questions for Further Investigation

Rather than settling the reincarnation question, current research has refined the issues that need addressing. If reincarnation represents genuine survival of consciousness, what mechanisms might enable the preservation and transmission of specific memories and characteristics? How could such mechanisms be detected and studied using scientific methods?

If reincarnation cases result from normal psychological and cultural processes, how can researchers better understand these processes and their effects? What kinds of controlled studies might distinguish between genuine survival and sophisticated false memory construction?

The development of new technologies offers potential approaches to these questions. Brain imaging techniques might reveal whether children's "past life memories" activate the same neural networks as genuine autobiographical memories. Genetic analysis could test whether apparent physical correspondences between children and deceased persons might result from unknown family connections.

More sophisticated statistical methods could help evaluate the significance of case correspondences while controlling for multiple comparisons and base rate effects. Cross-cultural studies could help distinguish universal features of the phenomenon from culture-specific interpretations and expectations.

Perhaps most importantly, the reincarnation debate highlights the need for better theoretical frameworks that can integrate findings from consciousness research, memory studies, child development, and cross-cultural psychology. Whether or not consciousness survives death, the phenomenon of children's past life claims reveals important aspects of how memory, identity, and cultural belief interact in human development.

The case of James Leininger continues to generate debate precisely because it combines elements that resist easy explanation: specific, verifiable details provided by a very young child, independent investigation by initially skeptical parents, and correspondences that extend beyond what coincidence would suggest. Whether this case represents evidence for survival or an illustration of how normal processes can create extraordinary-seeming results remains an open question—one that ultimately bears on fundamental questions about the nature of consciousness and the possibility of life beyond death.

Cultural Lens: The Interpretation Framework

The cross-cultural distribution of reincarnation cases raises crucial questions about the relationship between belief, culture, and reported experience. In societies where reincarnation is an accepted part of worldview—such as among Hindus, Buddhists, Druze, and certain Native American communities—children's unusual statements are readily interpreted as evidence of past lives and often become the focus of family and community investigation.

In contrast, Western families confronted with children's apparent past life memories often initially seek psychological or medical explanations, turning to reincarnation interpretations only when conventional approaches fail to explain their children's statements and behaviors. This cultural difference has profound implications for how cases develop and are documented.

The interpretive framework shapes not only how adults understand children's statements but potentially how children themselves experience and express their memories. A child growing up in a culture that expects and accepts reincarnation may feel free to elaborate on unusual memories or impressions, while a child in a skeptical culture may learn to suppress such expressions.

These cultural variations highlight the challenge of distinguishing universal features of the phenomenon from culture-specific interpretations. Are the apparent memories themselves shaped by cultural expectations, or do universal memory patterns simply receive different cultural interpretations? This question bears on fundamental issues in cross-cultural psychology and consciousness research.

Without careful attention to these interpretive frameworks, researchers risk either dismissing genuine cross-cultural patterns or mistaking cultural variations for evidence of universality. The challenge is to develop research approaches that can investigate the phenomenon while remaining sensitive to the cultural contexts in which it occurs.

Chapter 9: Dreams, Precognition, and Alternate Selves

On the evening of October 20, 1966, nine-year-old Eryl Mai Jones told her mother about a disturbing dream. "I dreamt I went to school and there was no school there," she said. "Something black had come down all over it." Her mother, Moya Jones, tried to comfort her daughter, but Eryl Mai remained troubled. "I'm not afraid to die," the child added. "I shall be with Peter and June"—two of her school friends.[1]

The next morning, October 21, Eryl Mai walked to Pantglas Junior School in Aberfan, Wales, as she did every day. At 9:15 AM, a mountain of coal waste that had been accumulating above the village for decades suddenly liquefied in the rain and crashed down the mountainside. The black avalanche buried the school, killing 144 people, including 116 children. Among the dead were Eryl Mai Jones, Peter, and June.[2]

In the weeks following the Aberfan disaster, psychiatrist John Barker of Shelton Hospital collected reports from people across Britain who claimed to have experienced premonitions of the tragedy. His systematic investigation, published in the *Journal of the Society for Psychical Research*, documented 76 claimed precognitive experiences, including 36 involving dreams. Many contained specific details that seemed to anticipate the disaster: black substances flowing down mountains, schools disappearing, children in danger.[3]

This tragedy became one of the most thoroughly documented cases of alleged mass precognition in modern history, raising profound questions about the nature of time, consciousness, and the possibility that human awareness might sometimes transcend the apparent boundaries of present-moment experience.

Defining Precognitive Dreams and Presentiment

Precognitive dreams represent a subset of anomalous temporal experiences where dream content appears to anticipate future events that could not have been predicted through normal inference or available information. Unlike ordinary dreams that process past experiences or current concerns, precognitive dreams allegedly contain specific details about events that have not yet occurred and that were unpredictable at the time of the dream.

Parapsychologist Louisa Rhine, who analyzed thousands of spontaneous cases reported to Duke University's Parapsychology Laboratory, identified several characteristic features of reported precognitive dreams. They typically involve emotionally significant events, often disasters or crises affecting the dreamer or their loved ones. The time interval between dream and fulfilling event is usually short—hours to days rather than weeks or months. The dreams often contain symbolic rather than literal imagery, requiring interpretation to recognize their apparent prophetic content.[4]

Related to precognitive dreams is the broader phenomenon of presentiment—subtle physiological changes that occur before unpredictable future events. Presentiment research, pioneered by researchers like Dean Radin and Dick Bierman, uses laboratory equipment to measure autonomic nervous system responses seconds before randomly selected emotional or neutral images are shown to subjects. Some studies report small but statistically significant differences in physiological arousal before emotional compared to neutral stimuli, suggesting that the body might somehow anticipate future events.[5]

The scientific investigation of these phenomena faces unique methodological challenges. Unlike other anomalous experiences that can be studied retrospectively, precognitive claims require prospective verification—the predicted event must actually occur for the experience to be considered successful. This creates a natural selection bias, as failed predictions are quickly forgotten while apparent hits are remembered and reported.

Cross-Cultural Patterns and Historical Context

Reports of prophetic dreams span virtually every culture and historical period, suggesting either a universal human experience or a universal human tendency to interpret unusual dreams in temporal terms. Ancient Mesopotamian dream texts describe elaborate procedures for incubating and interpreting prophetic dreams, while the Hebrew Bible contains numerous accounts of dreams that reveal future events.

In ancient Greece, the practice of dream incubation at temples dedicated to Asklepios involved sleeping in sacred spaces to receive divine guidance about future health or important decisions. The resulting dreams were interpreted by trained priests who specialized in understanding symbolic imagery and temporal significance.[6]

Indigenous Australian cultures understand dreaming as access to the Dreamtime—a timeless realm where past, present, and future coexist. In this worldview, prophetic dreams represent normal access to temporal information rather than anomalous experiences. Similar concepts appear in many Native American traditions, where vision quests and dream experiences are understood to provide guidance about future events and appropriate actions.[7]

Contemporary cross-cultural studies reveal interesting variations in both the reporting and interpretation of precognitive dreams. Cultures with strong beliefs in prophetic dreams report higher frequencies of such experiences, while cultures that emphasize linear time and materialist explanations report fewer cases. However, when precognitive dreams are reported in skeptical cultures, they often involve more specific details and more rigorous attempts at verification.

Anthropologist Barbara Tedlock's comparative study of dream practices found that cultures differ significantly in their criteria for recognizing prophetic dreams and their methods for interpreting temporal symbolism. Some cultures emphasize literal correspondences between dream imagery and future events, while others focus on symbolic or metaphorical connections that require skilled interpretation.[8]

The Scientific Debate: Three Distinct Positions

The investigation of precognitive dreams and presentiment has generated one of the most contentious debates in consciousness research, with three clearly defined positions that reflect fundamental disagreements about the nature of time, causality, and scientific evidence.

Physicalist Skeptics: Coincidence and Cognitive Bias

The mainstream scientific position, represented by researchers like Richard Wiseman, Susan Blackmore, and Christopher French, argues that precognitive dreams result entirely from well-understood psychological processes including coincidence, selective memory, and cognitive bias. There is no need to invoke anomalous temporal perception or retrocausal information transfer.

Wiseman, professor of psychology at the University of Hertfordshire, has conducted extensive research on the psychology of apparent psychic experiences. He argues that precognitive dreams represent a perfect example of how cognitive biases can create the illusion of paranormal phenomena. Given that people dream every night and that dreams often contain emotionally charged imagery involving disasters, accidents, or threats, it would be statistically inevitable that some dreams would appear to match subsequent real events purely by chance.[9]

When Wiseman analyzed the Aberfan case specifically, he noted several factors that could explain the apparent precognitive hits without invoking genuine prophesy. The coal tip above Aberfan had been recognized as dangerous for years, with previous smaller landslides occurring in 1944 and 1963. Local residents were aware of the potential hazard, making dreams about disasters in the area psychologically primed rather than genuinely prophetic. The black imagery reported in many dreams could reflect this existing knowledge and concern.[10]

Blackmore, formerly of the University of Plymouth, emphasizes the role of confirmation bias and selective reporting in creating apparent precognitive clusters around disasters like Aberfan. She points out that Barker's investigation began only after the disaster occurred, relying on people's retrospective reports of their dreams. Under these conditions, normal forgetting would eliminate dreams that didn't match the event while preserving and potentially embellishing those that seemed relevant.[11]

French, at Goldsmiths University of London, has studied the psychology of memory distortion in relation to apparent psychic experiences. His research shows how easily memories of dreams can be unconsciously modified after significant events occur, with people genuinely believing they had more specific or accurate premonitions than they actually experienced. The emotional impact of disasters like Aberfan could significantly enhance these memory distortion processes.[12]

When confronted with laboratory studies reporting statistical evidence for presentiment, physicalist skeptics emphasize methodological concerns and publication bias. They argue that the small effect sizes typically reported in presentiment studies are more likely to reflect subtle experimental artifacts than genuine retrocausal effects. The failure of many attempted replications suggests that initial positive results may have been statistical flukes rather than genuine discoveries.

Methodological Critics: Questioning Standards and Interpretation

A second group of researchers accepts that some reported precognitive experiences may be genuinely anomalous but raises substantial questions about research methodology and the interpretation of evidence. These critics don't necessarily reject the possibility of temporal anomalies but argue that current research methods are inadequate to demonstrate them convincingly.

Caroline Watt, professor of psychology at the University of Edinburgh and past president of the Parapsychological Association, exemplifies this position. Watt acknowledges that some precognitive cases present intriguing features that resist easy explanation but argues that the research methodology typically fails to meet rigorous scientific standards. The reliance on retrospective case collection, the lack of adequate controls for normal information sources, and the absence of reliable replication protocols all compromise the strength of the evidence.[13]

When methodological critics examine the Aberfan case, they note that Barker's investigation, while systematic for its time, lacked many features that would be required for contemporary research. No attempt was made to establish base rates for disaster-related dreams in the general population, making it impossible to determine whether the reported precognitive dreams occurred at frequencies exceeding chance expectations. The investigation also lacked independent verification of the original dream reports, relying primarily on the claimants' own accounts.

James Alcock, professor emeritus of psychology at York University, has conducted detailed analyses of precognitive dream research methodology. He points out that even apparently impressive cases often contain significant evidential gaps. Dream reports are typically vague enough to match multiple possible future events, and investigators may unconsciously engage in pattern-matching that emphasizes hits while overlooking misses. The lack of systematic protocols for recording and evaluating dream content makes objective assessment extremely difficult.[14]

Methodological critics also raise concerns about the laboratory studies of presentiment that some researchers cite as evidence for retrocausal effects. While these studies appear more controlled than spontaneous case investigations, they face their own methodological challenges. The effect sizes are typically very small, the replication rate is inconsistent, and subtle artifacts related to equipment function or data analysis could potentially account for the reported results.

However, methodological critics remain more open than physicalist skeptics to the possibility that genuine temporal anomalies might exist. They argue that better research designs, stronger controls, and more rigorous replication protocols might eventually provide convincing evidence for or against precognitive phenomena. The question is not whether such phenomena are theoretically impossible, but whether current evidence meets appropriate scientific standards.

Anomaly Proponents: Evidence for Temporal Flexibility

A third group of researchers, while maintaining scientific rigor, argues that accumulating evidence from precognitive dream research and laboratory presentiment studies points toward genuine temporal anomalies in consciousness that challenge conventional assumptions about causality and the nature of time.

Dean Radin, chief scientist at the Institute of Noetic Sciences, has conducted some of the most sophisticated laboratory investigations of presentiment effects. His research uses advanced physiological monitoring to measure subtle changes in skin conductance, heart rate, and brain activity seconds before randomly selected emotional images are presented to subjects. Meta-analyses of these studies, incorporating data from multiple laboratories, report small but statistically significant effects that appear to show the body responding to future events before they occur.[15]

Radin argues that these presentiment effects provide experimental evidence for retrocausal information transfer that could explain spontaneous precognitive experiences like those reported around Aberfan. While the laboratory effects are subtle—typically involving changes of a few percent in physiological measures—they are consistent enough across studies to suggest a genuine phenomenon rather than experimental artifact.

When anomaly proponents examine the Aberfan case, they emphasize features that resist conventional explanation. Several of the reported precognitive dreams contained specific details that were recorded before the disaster occurred, reducing the possibility of memory distortion. The geographic distribution of reports—coming from across Britain rather than just the local area—suggests something beyond local knowledge or concern about the coal tip's danger.

Daryl Bem, professor emeritus of psychology at Cornell University, conducted a series of controversial experiments published in the *Journal of Personality and Social Psychology* that appeared to demonstrate "feeling the future" effects in laboratory settings. His studies used variations of standard psychological paradigms but reversed the temporal order, testing whether future events could influence past responses. Several experiments reported statistically significant effects, suggesting that future random events could retroactively influence participants' earlier choices and responses.[16]

While Bem's studies generated significant controversy and criticism, they also inspired numerous replication attempts. Some independent laboratories reported successful replications of his effects, while others failed to find significant results. The mixed replication record has become a central focus of debate between anomaly proponents and their critics.

Julia Mossbridge, neuroscientist at Northwestern University, has conducted neuroimaging studies of presentiment effects using EEG and fMRI technology. Her research suggests that brain activity changes detectably before emotional compared to neutral stimuli are presented, with the most prominent effects occurring in regions associated with emotional processing and attention. These findings provide potential neural correlates for the physiological presentiment effects reported by other researchers.[17]

Anomaly proponents also point to the theoretical possibility of retrocausal effects within certain interpretations of quantum mechanics. While mainstream physics doesn't generally allow information to travel backward in time, some quantum theorists have proposed mechanisms that could permit retrocausal correlations under specific conditions. These theoretical frameworks remain highly speculative but provide conceptual foundations for understanding how precognitive phenomena might operate within natural law.

Frameworks from Part I: New Perspectives on Temporal Anomalies

The theoretical frameworks developed in Part I offer fresh approaches to understanding precognitive dreams and presentiment effects, though each comes with both explanatory insights and significant limitations when applied to the accumulated evidence.

The Predictive Brain: Sophisticated Forecasting or Temporal Access?

The predictive processing model provides perhaps the most compelling conventional framework for understanding apparent precognitive dreams. If the brain constantly generates internal models of likely future events based on available information, sleep states might allow these predictive mechanisms to operate with reduced reality-testing constraints, generating detailed scenarios that sometimes match subsequent events.

Memory researcher Lynn Hasher has shown how sleep states alter the normal filtering mechanisms that distinguish between internally generated thoughts and externally derived perceptions. During REM sleep, the brain's critical evaluation systems are largely offline, allowing dream content to feel as real and compelling as waking experience. This could explain why precognitive dreams often feel more vivid and meaningful than ordinary dreams.[18]

Applied to the Aberfan case, this framework suggests that local residents may have unconsciously processed multiple subtle cues about the coal tip's instability—previous small landslides, visible changes in the waste pile, weather patterns, geological knowledge—and integrated this information during sleep to generate disaster scenarios. The most emotionally compelling dreams would be remembered and reported when the predicted event actually occurred.

However, anomaly proponents argue that predictive processing models, while explaining many cases, cannot account for precognitive dreams that contain specific details apparently unavailable through normal information channels. Radin and his colleagues point to laboratory presentiment studies where the future events are determined by random number generators, making normal prediction impossible. If these effects are genuine, they suggest mechanisms beyond sophisticated biological forecasting.

Critics respond that even apparently random systems may contain subtle patterns that unconscious processing could detect. Computer-generated random sequences often contain hidden regularities, and laboratory equipment may introduce systematic biases that trained subjects could unconsciously learn to anticipate. The apparent specificity of some precognitive dreams may also reflect post-hoc pattern matching rather than genuine prophetic accuracy.

Consciousness Models: Linear Time or Eternal Now?

Theories about the fundamental nature of consciousness raise profound questions about temporal experience and the possibility of accessing information from outside the present moment. If consciousness is entirely emergent from neural activity operating within normal spacetime, then genuine precognition becomes essentially impossible. However, if consciousness involves fundamental features of reality that transcend local brain function, temporal anomalies become conceivable.

Philosopher Henri Bergson proposed that consciousness normally operates through a "reducing valve" that filters the vast potential awareness down to information relevant for immediate survival and action. In this view, ordinary consciousness is temporally constrained not because time travel is impossible, but because selective attention focuses awareness on present-moment concerns. Altered states like dreams might partially bypass this filtering, allowing access to temporally distributed information.[19]

Physicist Julian Barbour's "timeless" interpretation of physics suggests that past, present, and future may be equally real aspects of a static four-dimensional block universe. If consciousness could somehow access this block structure directly rather than experiencing it sequentially, precognitive phenomena would represent glimpses of the eternal "now" rather than violations of causality.[20]

However, neuroscientist Christof Koch argues that everything science has learned about consciousness points to its dependence on specific neural processes operating within normal temporal constraints. Memory formation, attention, and subjective experience all appear to require sequential information processing that operates according to standard causal principles. Without these temporal mechanisms, there would be no basis for coherent conscious experience, let alone precognitive access to future events.[21]

The debate ultimately confronts fundamental questions about the relationship between subjective time and physical time. Does consciousness create the experience of temporal flow, or does it simply detect temporal relationships that exist independently? If the former, precognitive dreams might represent anomalous states where normal temporal construction breaks down. If the latter, they might indicate access to temporal information that is normally filtered from awareness.

Quantum Mechanics and Retrocausality

Some researchers have proposed quantum mechanical processes as potential mechanisms for precognitive phenomena, drawing on interpretations of quantum theory that allow for retrocausal effects under specific conditions. While mainstream physics doesn't permit information transmission from future to past, certain quantum phenomena appear to involve correlations that transcend normal temporal ordering.

The quantum Zeno effect demonstrates that measurement can influence the evolution of quantum systems in ways that appear to involve backward causation. Similarly, the delayed-choice quantum eraser experiments show that decisions made in the future can appear to retroactively determine the behavior of particles in the past. While these effects operate at microscopic scales and don't permit macroscopic information transfer, they suggest that quantum mechanics allows more temporal flexibility than classical physics.[22]

Physicist Henry Stapp has proposed that consciousness might involve quantum measurement processes that could theoretically create retrocausal correlations between mental states and future events. In his interpretation, conscious observation collapses quantum wave functions in ways that could establish non-local connections across time as well as space.[23]

However, most quantum physicists remain deeply skeptical of such applications to macroscopic consciousness. Sean Carroll argues that quantum effects in biological systems decohere far too rapidly to support sustained retrocausal processes. The warm, wet environment of the brain would eliminate quantum coherence within femtoseconds, making quantum-based precognition effectively impossible.[24]

Max Tegmark's calculations suggest that quantum coherence in neural microtubules—the structures proposed by some theories of quantum consciousness—lasts for only about 10^{-13} seconds. This is far too brief to support the kind of sustained information processing that would be required for precognitive dreams or presentiment effects.[25]

Even researchers sympathetic to quantum approaches acknowledge that no known quantum mechanism could account for the specific, detailed information transfer that characterizes the strongest precognitive cases. Quantum correlations can create statistical anomalies but cannot transmit semantic content like the specific imagery reported in precognitive dreams.

Artificial Intelligence and Pattern Recognition

The development of sophisticated AI systems offers new metaphors for understanding how apparent precognitive experiences might arise through normal but highly sophisticated pattern recognition processes. Large language models can generate remarkably accurate predictions about likely future events based on pattern analysis of vast datasets, suggesting how biological intelligence might achieve similar predictive capabilities.

Cognitive scientist Andy Clark has proposed that human consciousness operates as a sophisticated prediction machine that constantly generates models of likely future states based on available information. These predictive models normally operate below conscious awareness, but sleep states might allow some predictions to surface as dream content that later appears prophetic when reality matches the projected scenarios.[26]

Applied to laboratory presentiment studies, this framework suggests that subjects might unconsciously detect subtle patterns in experimental procedures—equipment warming cycles, researcher behavior, environmental factors—that correlate with the timing of emotional stimulus presentation. The autonomic nervous system might respond to these detected patterns even when conscious awareness registers no relevant information.

However, anomaly proponents argue that this framework, while explaining how sophisticated prediction might occur, doesn't address studies where the future events are determined by quantum random number generators or other systems that should be genuinely unpredictable. If presentiment effects are real and involve truly random future events, they would require mechanisms beyond pattern recognition and statistical inference.

Critics respond that even apparently random systems often contain subtle regularities that sophisticated biological pattern recognition could potentially detect. The human nervous system evolved to detect extremely subtle environmental patterns related to survival threats, and modern experimental equipment may introduce systematic artifacts that unconscious processing could learn to anticipate.

Systems Theory and Temporal Networks

Network science and systems theory suggest ways that information about future events might propagate through complex systems in ways that create the appearance of precognitive access. In highly connected social networks, subtle information about impending events might spread through informal channels faster than official communication, reaching some individuals through unconscious perception before conscious awareness.

Applied to disaster cases like Aberfan, this framework suggests that multiple subtle indicators of impending catastrophe might propagate through the community via unconscious social cues, body language, and environmental observations. Sensitive individuals might integrate these distributed signals during sleep, generating dreams that appear prophetic but actually reflect sophisticated social information processing.

Biologist Rupert Sheldrake has proposed that morphic field effects might create information-sharing networks that transcend normal spatial and temporal boundaries. In this highly speculative framework, precognitive experiences might reflect access to information fields that connect conscious systems across time and space.[27]

However, mainstream biologists and physicists have found no evidence for morphic fields or similar non-local information networks. Controlled experiments designed to test morphic resonance have generally failed to produce significant results, and the proposed mechanisms conflict with well-established principles of physics and biology.

More conventionally, research on social contagion and collective behavior shows how information and emotional states can spread rapidly through connected populations via normal psychological and social mechanisms. These processes might account for some apparent collective precognitive experiences without requiring anomalous temporal access.

Sleep Research and Temporal Processing

Contemporary sleep and dream research provides important context for understanding how precognitive experiences might arise through normal but sophisticated neural processes operating during altered states of consciousness. REM sleep involves unique patterns of brain activity that differ significantly from both waking consciousness and other sleep stages.

Research by neuroscientist Matthew Walker has shown that sleep plays crucial roles in memory consolidation, pattern recognition, and creative problem-solving. During REM sleep, the brain reactivates and reorganizes experiences from the day, often combining them in novel ways that can generate insights not available during waking consciousness. This process might occasionally produce scenarios that appear prophetic when reality later matches the reorganized material.[28]

Studies of dream content reveal that dreams often contain temporal anomalies—past, present, and future elements combined in ways that violate normal chronological order. This temporal flexibility in dream construction might contribute to the feeling that some dreams transcend ordinary time boundaries, even when they result from normal memory and imagination processes.

However, research on dream recall and memory shows that dream content is highly susceptible to post-hoc modification. People often unconsciously alter their memories of dreams to better match subsequent events, genuinely believing they had more accurate premonitions than actually occurred. This memory plasticity significantly complicates the evaluation of spontaneous precognitive dream reports.

Sleep research also reveals individual differences in dream recall, emotional responsiveness, and pattern recognition that might influence the frequency of apparent precognitive experiences. People with high dream recall, strong emotional sensitivity, and sophisticated pattern recognition abilities might be more likely to generate and remember dreams that seem to anticipate future events.

Current State of Evidence: An Honest Assessment

After decades of research into precognitive dreams and laboratory presentiment studies, what can be concluded about the possibility of temporal anomalies in consciousness? The picture that emerges is complex and contentious, with different researchers drawing dramatically different conclusions from similar evidence.

The phenomenon of precognitive dreams is well-documented across cultures and historical periods. Collections like those compiled by Rhine and Barker contain thousands of reported cases, some with impressive details that were apparently recorded before the fulfilling events occurred. The consistency of reported features—emotional significance, short time intervals, symbolic content—suggests either genuine temporal anomalies or universal psychological processes that create similar illusions.

Laboratory studies of presentiment effects have produced a mixed but intriguing pattern of results. Meta-analyses of multiple studies report small but statistically significant effects suggesting that physiological measures can change before randomly selected emotional stimuli are presented. However, the effect sizes are small, replication attempts have yielded inconsistent results, and potential methodological artifacts remain incompletely understood.

The quality of precognitive dream research has improved significantly since early anecdotal collections. Modern investigations increasingly use prospective designs where dreams are recorded before verification attempts, independent witnesses document original reports, and systematic protocols attempt to rule out normal information sources. However, the fundamental methodological challenges of studying spontaneous temporal anomalies remain formidable.

Perhaps most importantly, no clear mechanism has been identified for how consciousness could access information about genuinely unpredictable future events. While various theoretical possibilities have been proposed—from quantum retrocausality to morphic field effects—none have gained sufficient empirical support to provide convincing explanations for the alleged phenomena.

The statistical evaluation of precognitive cases also presents ongoing challenges. Even apparently specific dream details may have higher base rates than initially apparent, and the number of potential matching events in any time period may be larger than investigators recognize. Without systematic studies comparing verified precognitive hits with false positives, it remains difficult to assess the true significance of apparent successes.

Cultural and psychological factors clearly influence both the reporting and interpretation of temporal anomalies. The overwhelming concentration of reports in cultures and individuals who believe in precognitive phenomena suggests either that belief affects occurrence or that it affects recognition and reporting. Both possibilities have important implications for how the evidence should be evaluated.

The Interpretive Challenge: Time, Causality, and Evidence

The precognitive dream debate ultimately confronts fundamental questions about the nature of time, causality, and the boundaries of scientific investigation. If consciousness sometimes transcends normal temporal constraints, it would require revising basic assumptions about the relationship between mind and physical reality. If precognitive experiences result entirely from sophisticated but conventional psychological processes, they reveal the extraordinary predictive capabilities of biological intelligence.

The debate also raises important questions about evidence standards and the burden of proof in science. How should unusual, well-documented cases be evaluated when they appear to challenge fundamental physical principles? What level of evidence should be required before considering explanations that contradict established scientific frameworks?

Skeptics argue that extraordinary claims require extraordinary evidence, and that normal explanations should always be preferred unless absolutely ruled out. They point to the long history of apparently paranormal phenomena that were eventually explained through conventional mechanisms, suggesting that precognitive dreams will yield to similar analysis given sufficient investigation.

Proponents respond that they are providing extraordinary evidence in the form of systematic case studies and controlled laboratory experiments. They argue that the consistent dismissal of evidence based on theoretical impossibility rather than empirical evaluation represents a form of scientific bias that impedes genuine inquiry into consciousness and temporality.

Methodological critics occupy a middle position, accepting that current evidence raises interesting questions while demanding higher standards of proof before accepting paradigm-challenging conclusions. They ask what kind of evidence would actually be convincing and whether current research methods can provide it.

The debate also highlights the challenge of studying subjective experiences that may involve fundamental features of consciousness not easily accessible to third-person scientific investigation. Precognitive dreams, if genuine, might represent aspects of temporal experience that require new methodological approaches integrating first-person and third-person perspectives.

Future Directions: Questions for Further Investigation

Rather than settling the question of temporal anomalies in consciousness, current research has identified specific issues that require further investigation. If precognitive experiences represent genuine retrocausal information transfer, what mechanisms might enable such processes while preserving causal consistency in the physical world? How could such mechanisms be detected and studied using scientific methods?

If precognitive dreams result from sophisticated biological forecasting and pattern recognition, how can researchers better understand these processes and their limitations? What kinds of information integration occur during sleep states, and how might they generate apparently prophetic content?

The development of new technologies offers potential approaches to these questions. Advanced neuroimaging might reveal whether precognitive dreams activate the same neural networks as memory recall or imagination, providing clues about their underlying mechanisms. Real-time dream monitoring could enable immediate recording of dream content before verification attempts, reducing memory distortion effects.

More sophisticated statistical methods could help evaluate the significance of apparent precognitive hits while controlling for multiple comparisons, base rate effects, and selection biases. Cross-cultural studies could help distinguish universal features of temporal experience from culture-specific interpretations and expectations.

Perhaps most importantly, the precognitive dream debate highlights the need for better theoretical frameworks that can integrate findings from consciousness research, sleep studies, memory research, and temporal physics. Whether or not consciousness can transcend normal temporal boundaries, the investigation of these phenomena reveals important aspects of how subjective time relates to physical time and how prediction and imagination operate in human cognition.

The case of Eryl Mai Jones and the Aberfan disaster continues to challenge researchers precisely because it combines elements that resist easy categorization: specific details apparently recorded before the event, emotional content consistent with the subsequent tragedy, and correspondence between dream imagery and actual disaster characteristics. Whether this case represents evidence for temporal anomalies or an illustration of how normal psychological processes can create extraordinary-seeming coincidences remains an open question—one that bears on fundamental issues about consciousness, time, and the nature of human experience.

Cultural Lens: Prophetic Dreams and Temporal Worldviews

The cross-cultural distribution and interpretation of precognitive dreams reveals fundamental differences in how various societies understand the nature of time and the relationship between consciousness and temporal reality. These cultural frameworks profoundly shape both the reported frequency of precognitive experiences and their interpretation within different knowledge systems.

In many indigenous cultures, dreams are understood as journeys through temporal landscapes where past, present, and future coexist simultaneously. The Aboriginal Australian concept of Dreamtime encompasses all temporal periods within a single accessible realm, making prophetic dreams a natural extension of ordinary dreaming rather than an anomalous experience requiring special explanation.

Western scientific culture, with its emphasis on linear causality and materialist explanations, tends to interpret precognitive dreams as either coincidences or evidence for unknown physical processes. This framework shapes both the kinds of questions researchers ask about temporal anomalies and the types of explanations they consider legitimate.

The contrast between these worldviews raises important questions about the relationship between cultural belief and reported experience. Do cultures with strong prophetic dream traditions experience more genuine temporal anomalies, or do they simply interpret ordinary psychological processes in temporal terms? How might Western scientific assumptions about time and causality limit the investigation of consciousness phenomena that transcend these frameworks?

These interpretive challenges extend beyond academic inquiry to practical questions about how different cultures understand and respond to apparent precognitive experiences. In societies that take prophetic dreams seriously, such experiences may influence important decisions about safety, relationships, and future planning. In skeptical cultures, similar experiences are more likely to be dismissed or pathologized, potentially overlooking both genuine insights and important psychological processes.

Chapter 10: Telepathy, Collective Minds, and Synchronicity

On March 14, 1979, identical twins Greta and Freda Chaplin were separated by several miles in Nottingham, England, when both suddenly experienced identical severe headaches at exactly 11:20 AM. Neither had been feeling unwell earlier. At the same moment, their mother, Barbara Chaplin, who was visiting Freda, felt compelled to telephone Greta immediately. When Greta answered, she was about to call Freda with the same concern. Both twins had experienced not only the same physical symptom but the same urgent need to contact each other. The headaches lasted exactly forty-seven minutes and ended simultaneously for both sisters.[1]

This case, documented by researcher Susan Blackmore during her investigation of twin telepathy claims, represents one of hundreds of reported instances where closely bonded individuals appear to share experiences across distance without any known means of communication. While skeptics attribute such events to coincidence and selective memory, the persistence of these reports across cultures and the emergence of laboratory studies showing statistical anomalies in information transfer have made telepathy one of the most systematically investigated—and contentious—phenomena in consciousness research.

Defining Telepathy and Related Phenomena

Telepathy, derived from the Greek words meaning "distant feeling," refers to the apparent direct transmission of thoughts, emotions, or sensory experiences between individuals without the mediation of known physical channels. The term was coined in 1882 by psychical researcher Frederic Myers, who sought to establish scientific frameworks for investigating what had previously been considered purely supernatural phenomena.[2]

Contemporary parapsychology distinguishes between several related but distinct categories of alleged information transfer. Crisis telepathy involves spontaneous awareness of distant emergencies or trauma affecting loved ones, often reported between family members or close friends during life-threatening situations. Experimental telepathy refers to controlled laboratory studies where subjects attempt to transmit or receive specific information under conditions designed to eliminate normal sensory cues.

Collective cognition represents a broader category encompassing apparent group intelligence that emerges from individual interactions but exceeds what would be expected from the sum of individual contributions. This includes phenomena observed in human groups, animal swarms, and distributed technological networks where coordinated behavior appears without centralized control or explicit communication.

Synchronicity, a concept introduced by Carl Jung in collaboration with physicist Wolfgang Pauli, describes meaningful coincidences that appear connected by significance rather than causation. Jung distinguished synchronicity from mere chance by its psychological impact and symbolic relevance to the individuals experiencing the coincidences.[3]

These phenomena share the suggestion that information or meaning can be transmitted or shared through channels that current scientific understanding cannot adequately explain, though they differ in their proposed mechanisms, empirical evidence, and theoretical implications.

Cross-Cultural Patterns and Historical Context

Reports of telepathic phenomena appear across virtually all human cultures, suggesting either universal psychological tendencies to interpret certain experiences in telepathic terms or genuine cross-cultural experiences that transcend normal communication channels. Australian Aboriginal cultures describe "yarning" connections that allow distant tribal members to share information through dreamtime experiences. Native American traditions include extensive accounts of medicine people receiving visions about distant events affecting their communities.[4]

In Hindu and Buddhist traditions, telepathic abilities (called "abhijna" in Sanskrit) are considered natural byproducts of advanced meditation practice. Classical yoga texts describe systematic methods for developing what they term "knowledge of others' minds," suggesting that some cultures have developed sophisticated frameworks for understanding and cultivating apparent telepathic abilities.[5]

African spiritual traditions often emphasize communal consciousness and the interconnectedness of minds within tribal or family groups. Anthropologist Richard Katz documented numerous accounts among the !Kung people of the Kalahari where healers claim to receive information about distant community members during trance states induced by traditional healing dances.[6]

In contemporary Western societies, telepathy reports are often associated with emotional crises, particularly between parents and children, spouses, or twins. The emotional intensity of these experiences appears to correlate with their reported frequency and vividness, suggesting that strong emotional bonds may somehow facilitate or enhance whatever mechanisms might underlie apparent telepathic phenomena.

However, cultural variations in both the interpretation and reporting of such experiences raise important questions about the relationship between belief systems and reported phenomena. Cultures that accept telepathy as normal may report higher frequencies of such experiences, while skeptical cultures may suppress or reinterpret similar events in more conventional terms.

The Scientific Investigation: From Anecdote to Experiment

The systematic study of telepathy began in the late 19th century with the founding of the Society for Psychical Research in London and similar organizations in other countries. Early investigators like Edmund Gurney and Frank Podmore collected thousands of spontaneous case reports and attempted to verify them through independent investigation.[7]

The transition from case studies to controlled experiments began in the 1930s with the work of Joseph Banks Rhine at Duke University. Rhine developed card-guessing experiments using specially designed symbols (ESP cards) to test whether subjects could identify targets selected by another person under controlled conditions. While Rhine's early results suggested above-chance performance, critics raised numerous methodological concerns about inadequate controls for sensory leakage and statistical analysis.[8]

The most sophisticated experimental approach to telepathy research has been the Ganzfeld procedure, developed in the 1970s by Charles Honorton and others. In Ganzfeld experiments, a "receiver" sits in sensory isolation with halved ping-pong balls over their eyes and white noise playing through headphones, while a "sender" in a separate room attempts to telepathically transmit randomly selected images or video clips. The receiver provides ongoing descriptions of mental imagery, which are later judged against the actual target and three decoy images.[9]

Meta-analyses of Ganzfeld studies spanning several decades have reported hit rates of approximately 32-35%, compared to the 25% expected by chance, yielding effect sizes that appear statistically significant but remain small and controversial. However, these results have been challenged by critics who point to methodological flaws, selective reporting, and the difficulty of achieving consistent replication across laboratories.

The Scientific Debate: Three Distinct Positions

The accumulated evidence from both spontaneous cases and laboratory studies has generated three clearly defined positions within the scientific community, each bringing different methodological assumptions and theoretical frameworks to bear on the telepathy question.

Mainstream Skeptics: No Psychic Information Transfer

The dominant position within mainstream science, represented by researchers like Ray Hyman, James Alcock, and Christopher French, argues that apparent telepathic phenomena result entirely from well-understood psychological processes including coincidence, selective memory, statistical artifacts, and subtle sensory cues that investigators fail to detect.

Ray Hyman, professor emeritus of psychology at the University of Oregon, has conducted some of the most detailed critiques of telepathy research methodology. Hyman argues that even the most sophisticated Ganzfeld studies contain subtle flaws that could account for the reported above-chance results. These include inadequate randomization of targets, insufficient isolation between sender and receiver, and unconscious bias in the judging process where evaluators may subtly favor matches between receiver descriptions and actual targets.[10]

Hyman's analysis of the Ganzfeld database reveals inconsistencies in effect sizes across different laboratories and investigators, suggesting that the results may reflect methodological artifacts rather than genuine telepathic phenomena. When studies with the most rigorous controls are analyzed separately, Hyman argues, the effect sizes drop toward chance levels, indicating that apparent positive results emerge primarily from less well-controlled experiments.

James Alcock, professor emeritus of psychology at York University, emphasizes the role of cognitive biases in creating the illusion of telepathic phenomena. Alcock points out that human beings are naturally pattern-seeking creatures who tend to notice and remember apparent hits while forgetting misses, creating the subjective impression of telepathic abilities even when no genuine information transfer occurs.[11]

When mainstream skeptics examine spontaneous cases like the Chaplin twins, they emphasize several factors that could create apparent telepathic experiences without invoking unknown mechanisms. Identical twins share genetic predispositions to similar health problems, making simultaneous symptoms more likely than initially apparent. The emotional bonds between twins may also create heightened attention to subtle behavioral cues that could trigger sympathetic responses.

Christopher French, at Goldsmiths University of London, has studied the psychology of belief in paranormal phenomena and argues that telepathy reports reflect a combination of coincidence, memory distortion, and the human tendency to find meaning in random events. French notes that the millions of potential telepathic "opportunities" that occur daily in a connected world make occasional striking coincidences statistically inevitable rather than evidentially significant.[12]

Experimental Critics: Questioning Methods and Standards

A second group of researchers accepts that some telepathy experiments may show genuine statistical anomalies but raises substantial questions about methodology, replication, and the interpretation of small effect sizes. These critics don't necessarily reject the possibility of anomalous information transfer but argue that current evidence fails to meet adequate scientific standards.

Jessica Utts, professor of statistics at the University of California, Irvine, represents this position in her ongoing debate with Ray Hyman about the interpretation of parapsychological evidence. Utts acknowledges methodological concerns in telepathy research but argues that the consistency of small positive effects across multiple laboratories and decades suggests that something genuinely anomalous may be occurring, even if the underlying mechanisms remain unclear.[13]

Utts points out that the effect sizes reported in telepathy research, while small, are comparable to those found in many accepted areas of psychology and medicine. The expectation that psychic phenomena should produce large, easily replicable effects may be unrealistic if such phenomena operate through subtle mechanisms that are easily disrupted by laboratory conditions.

Richard Wiseman, professor of psychology at the University of Hertfordshire, occupies a middle position between skepticism and acceptance. Wiseman has conducted his own telepathy experiments and found no evidence for psychic phenomena, but he acknowledges that some proponent researchers appear to obtain positive results that he cannot easily dismiss as fraudulent or incompetent.[14]

Wiseman suggests that the difference between "psychic" and "non-psychic" experimenters may reflect subtle psychological factors that influence subject performance in ways not yet understood. This "experimenter effect" could potentially explain both the positive results obtained by some researchers and the failures to replicate reported by others, without requiring either telepathic phenomena or methodological incompetence.

Caroline Watt, professor of psychology at the University of Edinburgh and former president of the Parapsychological Association, has studied both the methodology and the replication patterns in telepathy research. Watt notes that while some Ganzfeld studies do report positive results, the overall picture shows significant heterogeneity in effect sizes and success rates across laboratories, investigators, and time periods.[15]

Experimental critics argue that before telepathy can be accepted as genuine, researchers need to develop more reliable protocols that can consistently produce significant effects across different laboratories and investigators. The current pattern of occasional positive results interspersed with null findings suggests either that the phenomenon is extremely fragile and dependent on unknown factors, or that positive results reflect subtle methodological artifacts rather than genuine psychic abilities.

Anomaly Proponents: Evidence for Non-Local Information Transfer

A third group of researchers, while maintaining scientific rigor, argues that the accumulated evidence from telepathy research points toward genuine anomalous information transfer that challenges conventional assumptions about the boundaries of individual consciousness and the mechanisms of communication.

Dean Radin, chief scientist at the Institute of Noetic Sciences, has conducted meta-analyses of telepathy experiments spanning over a century of research. Radin argues that when all methodological concerns are taken into account, the cumulative evidence still shows effect sizes that exceed what would be expected from chance, fraud, or methodological artifacts alone. The consistency of small positive effects across different experimental paradigms, investigators, and historical periods suggests a genuine but subtle phenomenon.[16]

Radin points out that the expectation for large, easily replicable effects in psychic research may be misguided. Many accepted phenomena in physics and biology involve very small effect sizes that require sophisticated instrumentation and statistical analysis to detect. If telepathy operates through quantum-level or other subtle mechanisms, it might naturally produce the kind of small, statistically significant effects observed in the best studies.

Daryl Bem, professor emeritus of psychology at Cornell University, has applied rigorous experimental protocols to the study of telepathic and precognitive phenomena. Bem's studies, published in mainstream psychology journals, report statistically significant effects that appear to violate conventional assumptions about information transfer and temporal causation.[17]

When anomaly proponents address the replication difficulties that plague telepathy research, they argue that the phenomenon may be inherently dependent on psychological and environmental factors that are difficult to control in laboratory settings. Emotional state, belief systems, experimenter-subject rapport, and even geomagnetic activity have all been proposed as variables that might influence the manifestation of telepathic phenomena.

Rupert Sheldrake, former research fellow at Clare College, Cambridge, has developed theories of "morphic resonance" that could potentially provide mechanisms for telepathic information transfer. Sheldrake proposes that minds can access information through non-local fields that connect similar mental states across space and time, though his theories remain highly controversial and lack mainstream scientific acceptance.[18]

Proponents also emphasize that the methodological sophistication of contemporary telepathy research has evolved significantly from early anecdotal reports. Modern Ganzfeld studies employ automated randomization, computer-controlled stimulus presentation, and statistical protocols designed to eliminate human bias and sensory leakage. If these studies show genuine positive effects, proponents argue, they cannot be easily dismissed through appeals to methodological inadequacy.

Frameworks from Part I: New Perspectives on Information Transfer

The theoretical frameworks developed in Part I offer fresh approaches to understanding apparent telepathic phenomena, though each comes with both explanatory insights and significant limitations when applied to the accumulated evidence.

The Predictive Brain: Sophisticated Inference or Information Transfer?

The predictive processing model provides perhaps the most compelling conventional framework for understanding apparent telepathic experiences. If the brain constantly generates predictions about likely events based on subtle environmental cues, apparent telepathy might result from unconscious inference processes that operate below the threshold of conscious awareness.

Research by cognitive scientist Andy Clark suggests that human brains are extraordinarily sophisticated prediction machines that integrate vast amounts of subtle sensory information to anticipate future events and the mental states of others. Applied to telepathy reports, this framework suggests that apparent mind-reading might actually reflect sophisticated unconscious processing of micro-expressions, behavioral patterns, environmental cues, and statistical regularities.[19]

When applied to cases like the Chaplin twins, predictive processing models suggest that identical twins sharing genetic predispositions and environmental factors might unconsciously detect similar patterns of physiological change, leading to synchronized symptoms that appear telepathic but actually reflect shared biological rhythms and environmental sensitivities.

However, anomaly proponents argue that predictive processing models, while explaining many spontaneous cases, cannot account for controlled laboratory studies where normal sensory channels are eliminated and targets are selected through random processes. If Ganzfeld studies show genuine above-chance results under conditions of adequate sensory isolation, they suggest mechanisms beyond sophisticated biological prediction.

Critics respond that even apparently controlled laboratory conditions may contain subtle information leakage that sophisticated unconscious processing could detect. Computer randomization systems may contain hidden patterns, isolation chambers may not eliminate all sensory cues, and experimenters may unconsciously communicate information through body language or other subtle channels.

The predictive brain framework also struggles to explain the apparent selectivity of telepathic phenomena—why certain individuals or relationships seem more conducive to such experiences than others. If telepathy results from enhanced predictive processing, it should occur more randomly across the population rather than clustering around emotional bonds and specific psychological profiles.

Consciousness Models: Individual Minds or Collective Fields?

Theories about the fundamental nature of consciousness raise profound questions about the boundaries of individual awareness and the possibility of shared mental states. If consciousness emerges entirely from individual brain activity, genuine telepathy becomes essentially impossible. However, if consciousness involves fundamental features of reality that transcend local neural processing, telepathic phenomena become conceivable within expanded theoretical frameworks.

Philosopher David Chalmers has noted that the hard problem of consciousness—explaining how subjective experience arises from objective brain processes—leaves open possibilities that purely materialist approaches might miss. If consciousness involves non-physical properties or field-like characteristics, those properties might theoretically enable forms of information sharing that transcend normal spatial boundaries. [20]

Some interpretations of Integrated Information Theory, developed by Giulio Tononi, suggest that consciousness might involve information integration processes that could theoretically operate across multiple brains under specific conditions. While IIT doesn't explicitly predict telepathic phenomena, it provides frameworks for understanding how information integration might transcend individual neural boundaries. [21]

However, neuroscientist Christof Koch argues that everything science has learned about consciousness points to its dependence on specific neural processes operating within individual brains. The intricate timing, connectivity, and biochemical processes required for conscious experience appear to require the kind of precisely coordinated neural activity that can only occur within single nervous systems.[22]

Even if consciousness were somehow fundamental to reality, critics point out that telepathy would still require explaining how specific thoughts, emotions, or sensory experiences could be preserved and transmitted from one individual to another. The lack of any known mechanism for such information transfer remains a major obstacle for field-based theories of consciousness.

Quantum Mechanics and Non-Local Correlations

Some researchers have proposed quantum mechanical processes as potential mechanisms for telepathic phenomena, drawing on the non-local correlations demonstrated in quantum entanglement experiments. While quantum entanglement cannot transmit usable information faster than light, some theorists suggest that biological systems might utilize quantum effects in ways that could enable subtle forms of information correlation.

Physicist Henry Stapp has proposed that consciousness might involve quantum measurement processes that could create non-local connections between minds. In Stapp's interpretation, conscious observation collapses quantum wave functions in ways that could theoretically establish correlations between mental states across space, though the mechanisms remain highly speculative.[23]

Research on quantum effects in biological systems has revealed that quantum coherence plays important roles in photosynthesis, bird navigation, and possibly other biological processes. This suggests that living systems may utilize quantum phenomena more extensively than previously recognized, potentially providing foundations for quantum theories of consciousness and telepathy.[24]

However, most quantum physicists remain deeply skeptical of applications to macroscopic consciousness and telepathy. Sean Carroll argues that quantum effects in biological systems are typically limited to very specific molecular processes operating at extremely small scales and brief time periods. The warm, wet environment of the brain would cause quantum coherence to decohere within femtoseconds, making sustained quantum-based telepathy effectively impossible.[25]

Max Tegmark's calculations suggest that quantum coherence in neural structures lasts far too briefly to support the kind of sustained information processing required for telepathic communication. Even if quantum effects influence some aspects of neural function, they appear insufficient to account for the specific, meaningful information transfer that characterizes reported telepathic phenomena.[26]

Artificial Intelligence and Distributed Cognition

The development of distributed AI systems and swarm intelligence offers new metaphors for understanding how apparent collective cognition might emerge from individual interactions without requiring telepathic communication. Ant colonies, bird flocks, and distributed computer networks demonstrate sophisticated collective behaviors that arise from simple local interactions following basic rules.

Research by complexity scientist Craig Reynolds has shown how flocking behavior in birds emerges from individual agents following three simple rules: separation, alignment, and cohesion. No central control or telepathic communication is required to produce the remarkable coordination observed in murmurations of starlings or schools of fish.[27]

Applied to human groups, this framework suggests that apparent collective intelligence might emerge from unconscious behavioral synchronization, emotional contagion, and the amplification of subtle social cues rather than telepathic information transfer. Groups of humans may achieve coordination through rapid, unconscious processing of body language, vocal tones, pheromones, and other biological signals that operate below conscious awareness.

However, anomaly proponents argue that while distributed cognition models explain how coordination can emerge from local interactions, they don't address laboratory studies where individuals appear to access specific information that was never physically present in their environment. If Ganzfeld receivers can accurately describe randomly selected images that no one in their vicinity has seen, it suggests information transfer mechanisms beyond distributed biological processing.

Critics respond that even apparently isolated laboratory conditions may involve subtle information transmission through electromagnetic fields, vibrations, or other physical channels that investigators fail to detect. The development of increasingly sophisticated technology for detecting such information leakage remains an ongoing challenge in telepathy research.

Systems Theory and Synchronization

Network science reveals how complex systems can achieve remarkable synchronization through relatively simple coupling mechanisms. Research by Steven Strogatz has shown how populations of oscillators—from fireflies to heart pacemaker cells—can spontaneously synchronize their rhythms through weak mutual influences, creating coherent collective behaviors.[28]

Applied to human consciousness, this framework suggests that individuals in close proximity or emotional connection might unconsciously synchronize brain rhythms, autonomic nervous system activity, and behavioral patterns in ways that create the appearance of telepathic connection. Research has shown that people in conversation synchronize their brain waves, heart rates, and breathing patterns, potentially facilitating enhanced empathy and communication.[29]

Neuroscientist Uri Hasson's research on "brain coupling" during communication shows that successful communication involves the synchronization of neural activity between speakers and listeners. This neural coupling might provide mechanisms for enhanced empathy and understanding that could be misinterpreted as telepathic phenomena.[30]

However, while synchronization effects explain how individuals might develop enhanced mutual understanding and emotional resonance, they don't obviously account for the specific information transfer reported in telepathy studies. Synchronized brain rhythms might facilitate communication and empathy, but they don't clearly enable one person to access specific thoughts or sensory experiences of another.

Morphic field theories, proposed by Rupert Sheldrake, suggest that synchronization might operate through non-physical fields that connect similar mental states across space and time. While these theories provide potential mechanisms for telepathic phenomena, they lack empirical support and contradict well-established principles of physics and biology.

Current State of Evidence: An Honest Assessment

After more than a century of systematic investigation, what can be concluded about telepathy and related phenomena? The picture that emerges is complex and contentious, with different researchers drawing dramatically different conclusions from similar evidence.

The phenomenon of spontaneous telepathic experiences is well-documented across cultures and historical periods. Collections of case studies contain thousands of reported instances, some with impressive details and independent verification. The consistency of certain features—emotional significance, close relationships, crisis situations—suggests either genuine information transfer mechanisms or universal psychological processes that create similar experiences.

Laboratory studies of telepathy have produced mixed but intriguing results. Meta-analyses of Ganzfeld experiments spanning several decades report small but statistically significant effect sizes, while individual studies show considerable variability in success rates. The overall pattern suggests either a genuine but fragile phenomenon or systematic methodological artifacts that create the appearance of positive results.

The quality of telepathy research has improved significantly since early anecdotal investigations. Modern studies employ sophisticated randomization procedures, computer-controlled stimulus presentation, and statistical protocols designed to eliminate human bias and sensory leakage. However, fundamental methodological challenges remain, particularly in achieving consistent replication across different laboratories and investigators.

Perhaps most importantly, no clear mechanism has been identified for how specific thoughts or experiences could be directly transmitted from one mind to another without the mediation of known physical processes. While various theoretical possibilities have been proposed—from quantum field effects to morphic resonance—none have gained sufficient empirical support to provide convincing explanations for the alleged phenomena.

The statistical evaluation of telepathy research also presents ongoing challenges. Even small effect sizes can achieve statistical significance when accumulated across large numbers of trials, but determining whether such effects reflect genuine phenomena or subtle methodological artifacts remains difficult.

The expectation that telepathic phenomena should produce large, obvious effects may be unrealistic if such phenomena operate through mechanisms that are naturally subtle and easily disrupted.

Collective Intelligence: From Swarms to Human Groups

The investigation of apparent collective cognition in human groups provides an interesting parallel to telepathy research, offering examples of genuine emergent intelligence that emerges from individual interactions without requiring telepathic communication. Research on swarm intelligence, crowd wisdom, and collective problem-solving reveals that groups can often outperform individuals through purely conventional information aggregation and processing mechanisms.

Studies of prediction markets, Wikipedia editing, and scientific collaboration show how large groups can achieve remarkable accuracy and creativity through the combination of diverse individual contributions. These examples of genuine collective intelligence provide important contrasts to alleged telepathic phenomena, demonstrating how apparent group minds can emerge through transparent, well-understood mechanisms.

However, some reported instances of human collective cognition appear to exceed what would be expected from conventional information aggregation. Cases where groups achieve coordination or solve problems with apparent access to information that was not explicitly shared among members continue to challenge researchers seeking purely conventional explanations.

The boundary between genuine collective intelligence and apparent telepathic phenomena remains unclear, with important implications for understanding both individual consciousness and group dynamics. Future research may reveal that these phenomena exist on a continuum rather than representing distinct categories of experience.

The Synchronicity Question: Meaning Without Causation

Carl Jung's concept of synchronicity presents unique challenges for scientific investigation because it involves subjective assessments of meaning and significance that resist objective measurement. While coincidences can be statistically analyzed, determining whether specific coincidences are "meaningful" requires psychological and cultural judgments that vary among individuals and societies.

Research on the psychology of coincidence perception reveals that humans are naturally predisposed to find meaningful patterns in random events, particularly during periods of emotional stress or major life transitions. This suggests that many experiences of synchronicity might reflect normal psychological processes rather than genuine acausal connections between inner and outer events.

However, some reported synchronicities involve such improbable coincidences that pure chance seems inadequate as an explanation. Jung's own examples often involved precise correspondences between psychological states and external events that occurred with timing that appears to exceed reasonable probability thresholds.

The challenge for scientific investigation is developing methods that can distinguish between psychologically meaningful coincidences that reflect normal pattern-seeking tendencies and potentially genuine instances of acausal connection. This requires integrating objective statistical analysis with subjective assessments of personal meaning and significance.

Future Directions: Questions for Further Investigation

Rather than settling questions about telepathy and collective consciousness, current research has identified specific issues that require further investigation. If telepathic phenomena represent genuine information transfer, what mechanisms might enable such processes while remaining consistent with known physical laws? How could such mechanisms be detected and studied using scientific methods?

If apparent telepathic experiences result from sophisticated unconscious inference and social coordination, how can researchers better understand these processes and their limitations? What kinds of subtle information exchange occur in human social interactions, and how might they create the appearance of telepathic communication?

The development of new technologies offers potential approaches to these questions. Advanced neuroimaging might reveal whether telepathic experiences activate the same neural networks as normal communication or represent distinct patterns of brain activity. Real-time monitoring of multiple brains during telepathy experiments could help identify whether genuine information transfer occurs or whether apparent correlations reflect synchronized responses to shared stimuli.

More sophisticated experimental designs could help isolate genuine telepathic effects from methodological artifacts. The use of multiple independent laboratories, automated experimental procedures, and advanced statistical techniques might help determine whether positive results in telepathy research reflect genuine phenomena or subtle experimental biases.

Perhaps most importantly, the investigation of telepathy and collective consciousness highlights the need for better theoretical frameworks that can integrate findings from consciousness research, social psychology, network science, and information theory. Whether or not minds can directly communicate across space, the study of these phenomena reveals important aspects of human social cognition, empathy, and the boundaries of individual awareness.

The case of the Chaplin twins continues to puzzle researchers precisely because it combines elements that resist easy categorization: precise timing of shared symptoms, emotional synchronization across distance, and behavioral coordination that appeared to exceed normal communication channels. Whether this case represents evidence for telepathic phenomena or an illustration of how normal biological and psychological processes can create extraordinary-seeming coincidences remains an open question—one that bears on fundamental issues about consciousness, communication, and the nature of human social connection.

Cultural Lens: From Psychic Powers to Social Networks

The cross-cultural interpretation of telepathic phenomena reveals fundamental differences in how various societies understand individual consciousness and social connection. These cultural frameworks profoundly shape both the reported frequency of telepathic experiences and their interpretation within different knowledge systems.

Western scientific culture, with its emphasis on individual consciousness and mechanistic causation, tends to interpret telepathic reports as either evidence for unknown physical forces or products of psychological bias and coincidence. This framework shapes both the kinds of questions researchers ask about information transfer and the types of explanations they consider legitimate.

Many non-Western cultures view individual consciousness as inherently connected to collective awareness, making telepathic experiences natural extensions of normal social cognition rather than anomalous phenomena requiring special explanation. In these contexts, the apparent communication between minds is understood as evidence for the fundamental interconnectedness of consciousness rather than violations of physical law.

Indigenous cultures often emphasize the continuity between human consciousness and natural systems, viewing telepathic experiences as part of broader ecological communication networks that include animals, plants, and landscape features. These worldviews provide interpretive frameworks that normalize rather than problematize apparent non-local information transfer.

The contrast between individualistic and collectivistic cultural orientations raises important questions about the relationship between social organization and reported telepathic phenomena. Do cultures that emphasize collective identity and social interdependence experience more genuine telepathic phenomena, or do they simply interpret normal social psychological processes in telepathic terms?

These interpretive challenges extend beyond academic research to practical questions about how different cultures understand and utilize apparent telepathic abilities. In societies that take such phenomena seriously, telepathic experiences may influence important decisions about relationships, community safety, and social coordination. In skeptical cultures, similar experiences are more likely to be dismissed or pathologized, potentially overlooking both genuine insights and important social psychological processes.

Chapter 11: Psychedelic Encounters and Non-Human Intelligences

On a Tuesday morning in December 1990, "Sarah," a 39-year-old psychologist, entered the research unit at the University of New Mexico Hospital. She had volunteered for Dr. Rick Strassman's groundbreaking study of DMT (dimethyltryptamine), becoming the first person to receive intravenous administration of this powerful psychedelic under controlled scientific conditions. Within seconds of the injection, Sarah's consciousness was catapulted into what she later described as "a place more real than this room."[1]

In her carefully documented report, Sarah described encountering "two beings that were definitely there, definitely separate from me, definitely real." These entities appeared as luminous, geometric forms that seemed to possess vast intelligence and communicated without words. They showed her complex visual patterns that felt like lessons about the nature of reality itself. When the fifteen-minute experience ended, Sarah was adamant that she had not hallucinated but had genuinely traveled to another realm and encountered autonomous intelligences.[2]

This account, one of hundreds documented in Strassman's pioneering research, launched a new era of scientific investigation into one of the most extraordinary and controversial aspects of human consciousness: the apparent ability of psychedelic substances to facilitate encounters with what seem to be non-human intelligences. These experiences challenge fundamental assumptions about the nature of consciousness, reality, and the boundaries of the self.

Defining Psychedelic Entity Encounters

Psychedelic entity encounters represent a distinct category of altered state experiences characterized by the perception of meeting conscious beings that appear separate from the self and possess apparent autonomy, intelligence, and intentionality. Unlike other types of hallucinations or mental imagery, these encounters typically involve interactive communication, often including the transmission of complex information that subjects feel unable to generate from their own knowledge or imagination.

These experiences occur across different psychedelic substances but show particularly high frequency with DMT, often called "the spirit molecule" for its propensity to induce such encounters. Research by psychiatrist Rick Strassman found that approximately 50% of his DMT study participants reported contact with non-human entities, making this one of the most common features of the DMT experience.[3]

The entities encountered vary widely in appearance and behavior but fall into several recurring categories. "Machine elves," a term popularized by philosopher Terence McKenna, appear as technological, often geometric beings that seem to inhabit spaces resembling advanced laboratories or workshops. "Insectoid beings" resemble intelligent insects or arachnids, often described as conducting examinations or procedures on the experiencer. "Light beings" appear as luminous, sometimes angelic presences that communicate through direct transmission of knowledge or emotion.[4]

Contemporary researchers distinguish entity encounters from other psychedelic phenomena such as geometric visual patterns, synesthesia, or ego dissolution. While these other effects are clearly recognized as alterations in perception and consciousness, entity encounters possess what philosophers call "otherness"—a quality of seeming to originate from sources external to the experiencer's own mind.

The phenomenology of these encounters shares striking similarities with reported alien abduction experiences, though they occur in very different contexts and typically lack the fear and trauma associated with abduction accounts. They also overlap with historical reports of encounters with spirits, angels, demons, and other non-human intelligences across various cultural and religious traditions.

Cross-Cultural Patterns and Indigenous Perspectives

Reports of encounters with non-human intelligences during altered states appear across virtually all cultures that use psychoactive plants for spiritual or healing purposes, suggesting either universal neurological phenomena or genuine contact with transpersonal realms that different cultures interpret through their own symbolic frameworks.

In the Amazon basin, where ayahuasca has been used for centuries by indigenous peoples, encounters with plant spirits are considered fundamental aspects of healing and spiritual development. The Shipibo-Conibo people of Peru describe detailed interactions with plant teachers who provide knowledge about medicine, ecology, and cosmic order. These beings are often visualized as serpents, jaguars, or hybrid human-plant entities that communicate through complex geometric patterns called "icaros."[5]

Anthropologist Luis Eduardo Luna's extensive research with Amazonian shamans reveals sophisticated taxonomies of spiritual entities that are contacted through ayahuasca ceremonies. These include "plant teachers" associated with specific medicinal plants, "forest spirits" that govern particular ecological domains, and "cosmic entities" that provide knowledge about fundamental principles of reality. Shamans describe these encounters as literal meetings with autonomous intelligences rather than symbolic or metaphorical experiences.[6]

In North American indigenous traditions, peyote ceremonies often involve encounters with spiritual beings that provide guidance, healing, and visionary knowledge. The Native American Church describes meetings with "Peyote Spirit," often visualized as a beautiful woman or wise elder who teaches about proper living and spiritual development. These encounters are understood as genuine contact with sacred presences rather than drug-induced hallucinations.[7]

Tibetan Buddhist traditions include sophisticated frameworks for understanding encounters with non-human intelligences during meditation and ritual practices. The *Bardo Thodol* (Tibetan Book of the Dead) describes detailed encounters with various categories of beings—peaceful deities, wrathful deities, hungry ghosts, and other entities—that consciousness may encounter during transitions between life and death. While these traditions don't typically use psychedelic substances, the phenomenology shows remarkable similarities to contemporary entity encounter reports.[8]

African spiritual traditions often involve encounters with ancestral spirits and nature deities during trance states induced by drumming, dancing, and sometimes plant medicines. The Bwiti religion of Gabon uses iboga root to facilitate encounters with ancestral spirits who provide guidance about personal and community problems. Practitioners describe these meetings as literal conversations with deceased relatives and spiritual guardians.[9]

However, Western secular cultures typically interpret entity encounters as products of brain chemistry rather than genuine contact with external intelligences. This interpretive difference raises profound questions about the relationship between cultural worldview and reported experience. Do cultures that accept the reality of spiritual entities actually facilitate genuine contact with such beings, or do they simply provide interpretive frameworks that make neurochemical hallucinations seem real?

The Scientific Investigation: From Phenomenology to Neuroscience

The systematic study of psychedelic entity encounters began in earnest in the 1990s with Strassman's DMT research at the University of New Mexico. Using rigorous medical protocols and detailed interviews, Strassman documented hundreds of entity encounter reports from volunteers who received intravenous DMT under controlled laboratory conditions. His work marked the first time such experiences had been systematically studied in a university research setting.[10]

Contemporary research has expanded to include neuroimaging studies that reveal the brain changes associated with psychedelic experiences. Robin Carhart-Harris and his colleagues at Imperial College London have used fMRI and other imaging techniques to study the neural correlates of psilocybin, LSD, and DMT experiences, including entity encounters. Their research reveals that psychedelics cause widespread changes in brain connectivity and activity patterns that may explain some features of these extraordinary experiences.[11]

Studies using the 5-Dimensional Altered States of Consciousness Rating Scale have found that entity encounters correlate with specific clusters of psychological effects including oceanic boundlessness, visual restructuralization, and elementary hallucinations. However, the encounters also show unique features that distinguish them from other types of psychedelic phenomena, suggesting they may involve distinct neurological mechanisms.[12]

Research on the therapeutic applications of psychedelics has revealed that entity encounters often play important roles in healing and personal transformation. Studies of psilocybin-assisted therapy for depression and PTSD report that many patients describe receiving guidance or healing from entities encountered during their sessions. Whether these beings are "real" or not, their apparent advice often proves psychologically beneficial and clinically relevant.[13]

The Scientific Debate: Three Distinct Interpretive Positions

The investigation of psychedelic entity encounters has generated intense debate within neuroscience, psychology, and consciousness studies, with three clearly defined positions emerging from the accumulating research evidence.

Neurobiological Reductionists: Entities as Neural Hallucinations

The mainstream scientific position, represented by researchers like Susan Greenfield, Michael Persinger, and most contemporary neuroscientists, argues that entity encounters result entirely from known neurochemical processes that create vivid but ultimately illusory experiences of contact with external intelligences.

Susan Greenfield, professor of pharmacology at Oxford University, argues that psychedelic substances disrupt normal brain function in ways that create compelling but false perceptions of external entities. Greenfield emphasizes that consciousness emerges from complex neural networks, and psychedelics alter these networks in ways that can generate experiences that feel completely real but correspond to no external reality.[14]

Greenfield points out that the specific content of entity encounters often reflects cultural expectations and personal psychology rather than contact with genuinely external beings. Western users frequently report technological or alien-like entities that reflect contemporary cultural anxieties about technology and space exploration, while indigenous users more often encounter animal spirits and plant teachers that align with their traditional cosmologies.

Michael Persinger, professor of psychology at Laurentian University, has conducted extensive research on the neurological basis of spiritual and anomalous experiences. Persinger argues that entity encounters result from abnormal activity in the temporal lobes, which are involved in self-other discrimination and social cognition. Psychedelics may stimulate these brain regions in ways that create false perceptions of external presence and agency.[15]

When neurobiological reductionists examine Strassman's DMT research, they emphasize that the experiences occurred under conditions of known neurochemical disruption. DMT is a powerful serotonin receptor agonist that dramatically alters neurotransmitter function throughout the brain. The resulting experiences, however compelling subjectively, reflect the brain's attempt to make sense of severely disrupted neural processing rather than genuine contact with external entities.

Robin Carhart-Harris, despite his pioneering research on psychedelic neuroscience, maintains a cautious position regarding the ontological status of entity encounters. His "entropic brain" hypothesis suggests that psychedelics reduce the brain's predictive processing capabilities, allowing normally suppressed neural activity to reach consciousness. Entity encounters may result from the interpretation of this disinhibited neural activity as external presence.[16]

Neurobiological reductionists also point to the consistency of entity encounters across different subjects as evidence for shared neurobiological mechanisms rather than contact with genuine external beings. If these experiences reflected real encounters with autonomous entities, they argue, we would expect much greater variability in reports rather than the recurring patterns observed across different users and cultures.

Phenomenological Agnostics: Taking Experience Seriously Without Ontological Commitment

A second group of researchers, while remaining within scientific frameworks, argues that entity encounters represent genuine psychological phenomena that deserve serious investigation regardless of their ultimate ontological status. These researchers focus on the phenomenology and effects of such experiences without making claims about their relationship to external reality.

David Luke, associate professor of psychology at the University of Greenwich, represents this position in his extensive research on psychedelic entity encounters. Luke argues that whether or not the entities are "real" in an objective sense, they represent consistent features of human consciousness under specific pharmacological conditions that have important psychological and anthropological significance.[17]

Luke's research reveals that entity encounters often involve the transmission of information that subjects report as novel and valuable, including insights about personal problems, creative solutions, and sometimes even technical knowledge that seems beyond the experiencer's normal capabilities. While this could reflect enhanced access to unconscious knowledge rather than contact with external intelligences, the phenomenon deserves serious scientific attention.

Christopher Timmermann, a researcher at Imperial College London, has conducted some of the most detailed contemporary studies of DMT entity encounters using advanced neuroimaging techniques. Timmermann argues that these experiences represent important windows into the nature of consciousness and social cognition that shouldn't be dismissed simply because they occur under pharmacologically altered conditions.[18]

Phenomenological agnostics emphasize that entity encounters have measurable psychological effects that persist long after the acute drug effects have ended. Studies of ayahuasca users report lasting changes in personality, worldview, and psychological well-being that often relate directly to guidance received from entities encountered during ceremonies. These effects suggest that the encounters, whatever their ultimate nature, engage fundamental aspects of human psychology and meaning-making.

Benny Shanon, professor of psychology at the Hebrew University of Jerusalem, has conducted extensive phenomenological research on ayahuasca experiences, including detailed analysis of entity encounters. Shanon argues that these experiences reveal important aspects of human consciousness that are normally hidden but become accessible under specific pharmacological conditions.[19]

When phenomenological agnostics examine claims about information transmission from entities, they note that such reports require careful investigation rather than immediate dismissal. While extraordinary claims require extraordinary evidence, the consistent reports of novel information transmission deserve systematic study rather than a priori rejection based on materialist assumptions.

Ontological Realists: Entities as Genuine Non-Human Intelligences

A third group of researchers, while maintaining scientific methods, argues that accumulating evidence suggests entity encounters may involve genuine contact with autonomous intelligences that exist independently of human consciousness. These researchers don't necessarily embrace supernatural explanations but contend that current materialist models may be insufficient to explain all features of these experiences.

Rick Strassman, the pioneering DMT researcher, has gradually moved toward more open interpretations of entity encounters based on his extensive documentation of such experiences. Strassman argues that the consistency, interactivity, and apparent autonomy of the entities encountered by his research subjects suggest something more than simple hallucinations or psychological projections.[20]

Strassman points out that many of his subjects were materialistic scientists and healthcare professionals who had no prior beliefs in spiritual entities or non-physical realms. Yet these individuals often emerged from their DMT experiences convinced that they had encountered genuinely autonomous beings possessing knowledge and intentions independent of their own minds.

Graham Hancock, while not a traditional academic researcher, has compiled extensive documentation of entity encounters across different psychedelic substances and cultural contexts. Hancock argues that the remarkable consistency of certain features—particularly the entities' apparent knowledge of human affairs and their tendency to communicate important information—suggests contact with genuine non-human intelligences rather than mere hallucinations.[21]

Stanislav Grof, psychiatrist and consciousness researcher, has documented thousands of psychedelic sessions over several decades of clinical work. Grof argues that entity encounters represent access to transpersonal domains of consciousness that include autonomous intelligences existing independently of individual human minds. His transpersonal psychology framework provides theoretical foundations for understanding such encounters as genuine contact with non-local consciousness.[22]

When ontological realists address skeptical challenges about the mechanism of such contact, they often point to quantum theories of consciousness and speculative physics models that could theoretically allow interaction with non-physical intelligences. While these theories remain highly speculative, they provide conceptual frameworks within which entity encounters could represent genuine rather than illusory experiences.

Dennis McKenna, ethnopharmacologist and brother of Terence McKenna, argues that psychedelic substances may function as "technologies" for accessing information domains that exist independently of individual brains. McKenna suggests that the entities encountered represent genuine inhabitants of these domains rather than projections of human psychology.[23]

Ontological realists also emphasize cases where entities have allegedly provided information that was subsequently verified as accurate but was not previously known to the experiencer. While such cases remain anecdotal and require rigorous investigation, they suggest possibilities beyond simple hallucination or psychological projection.

Frameworks from Part I: New Lenses for Understanding Entity Encounters

The theoretical frameworks developed in Part I offer diverse approaches to understanding psychedelic entity encounters, though each comes with both explanatory insights and significant limitations when applied to these extraordinary experiences.

The Predictive Brain: Disrupted Processing and False Entities

The predictive processing model provides perhaps the most compelling conventional framework for understanding entity encounters during psychedelic experiences. If the brain constantly generates predictions about sensory input and social encounters, psychedelic substances might disrupt these predictive mechanisms in ways that create convincing but false perceptions of external entities.

Research by Andy Clark and Jakob Hohwy suggests that consciousness emerges from the brain's predictive models of both external environments and internal states. Psychedelics appear to reduce the influence of high-level predictive models while amplifying bottom-up sensory processing, potentially creating conditions where the brain interprets random neural activity as evidence of external agency and intelligence.[24]

Applied to entity encounters, this framework suggests that psychedelics might disinhibit social cognition networks that normally detect and interpret human social cues, causing these systems to interpret ambiguous internal signals as evidence of external presence. The apparent intelligence and autonomy of encountered entities might reflect the sophistication of human social cognitive mechanisms rather than genuine contact with external beings.

The model also explains why entity encounters often involve communication and information transmission. Human brains are exquisitely sensitive to social interaction patterns, and the same mechanisms that allow us to interpret facial expressions, voice tones, and body language in normal social encounters might generate compelling experiences of communication when applied to psychedelically-induced neural activity.

However, anomaly proponents argue that predictive processing models, while explaining many features of entity encounters, struggle to account for cases where entities allegedly provide information that was not previously accessible to the experiencer's brain. If the encounters represent purely internal brain processes, how could they generate genuinely novel knowledge about external reality?

Critics respond that apparent information transmission might reflect enhanced access to unconscious knowledge, pattern recognition, or creative problem-solving abilities rather than communication with external entities. Psychedelics might temporarily enhance cognitive abilities in ways that feel like receiving information from external sources but actually reflect the brain's own enhanced processing capabilities.

Consciousness Models: Individual Minds or Shared Reality?

Theories about the fundamental nature of consciousness raise profound questions about the boundaries of individual awareness and the possibility of encountering other forms of consciousness during altered states. If consciousness emerges entirely from individual brain activity, genuine entity encounters become essentially impossible. However, if consciousness involves fundamental features of reality that transcend individual neural processing, such encounters become conceivable within expanded theoretical frameworks.

David Chalmers' analysis of the hard problem of consciousness—explaining how subjective experience arises from objective brain processes—leaves open possibilities that purely materialist approaches might miss. If consciousness involves non-physical properties or participates in fundamental features of reality, those properties might theoretically enable encounters with other forms of consciousness during states where normal cognitive boundaries are altered.[25]

Some interpretations of Integrated Information Theory, developed by Giulio Tononi, suggest that consciousness involves information integration processes that could theoretically operate across multiple systems under specific conditions. While IIT doesn't predict entity encounters, it provides frameworks for understanding how consciousness might interact with other conscious systems in ways that transcend normal spatial and temporal boundaries.[26]

Philosopher Philip Goff's panpsychist theories propose that consciousness represents a fundamental feature of reality that exists at multiple scales and in various forms. In this framework, psychedelic entity encounters might represent temporary access to forms of consciousness that exist independently of human brains but are normally filtered from awareness by cognitive mechanisms focused on survival and social functioning.[27]

However, neuroscientist Christof Koch argues that everything science has learned about consciousness points to its dependence on specific neural processes operating within individual brains. The intricate timing, connectivity, and biochemical processes required for conscious experience appear to require the kind of precisely coordinated neural activity that can only occur within single nervous systems.[28]

Even if consciousness were somehow fundamental to reality, critics point out that entity encounters would still require explaining how specific forms of intelligence could exist independently of biological or technological substrates and how psychedelic substances could enable contact with such entities. The lack of any known mechanism for such interaction remains a major obstacle for consciousness-based theories of entity encounters.

Quantum Mechanics and Information Fields

Some researchers have proposed quantum mechanical processes as potential mechanisms for psychedelic entity encounters, drawing on quantum field theories and speculative physics models that could theoretically enable interaction with non-physical forms of consciousness or information.

Henry Stapp's quantum theories of consciousness suggest that conscious observation involves quantum measurement processes that could theoretically create non-local connections between different conscious systems. In this framework, psychedelic substances might alter consciousness in ways that enable quantum-level interactions with other forms of conscious experience.[29]

Stuart Hameroff and Roger Penrose's Orchestrated Objective Reduction theory proposes that consciousness emerges from quantum processes in neural microtubules. While their theory primarily addresses consciousness in living brains, Hameroff has speculated that quantum information might exist in forms that could theoretically interact with altered human consciousness during psychedelic experiences.[30]

Some theoretical physicists have proposed models of reality that include multiple dimensions or parallel universes that could theoretically harbor forms of consciousness different from those found in our normal three-dimensional space. Psychedelic entity encounters might represent temporary access to such alternate dimensional spaces or contact with consciousness existing in parallel realities.

However, most quantum physicists remain deeply skeptical of such applications to macroscopic consciousness and entity encounters. Sean Carroll argues that quantum effects in biological systems are typically limited to very specific molecular processes that operate at extremely small scales and brief time periods. The warm, wet environment of the brain would cause quantum coherence to decohere within femtoseconds, making sustained quantum-based entity contact effectively impossible.[31]

Even if quantum effects could somehow enable contact with non-physical entities, critics note that such theories provide no specific mechanisms for how meaningful communication could occur or how entities existing in quantum fields could possess the kind of complex intelligence and knowledge attributed to them in encounter reports.

Artificial Intelligence and Emergent Agency

The development of sophisticated AI systems offers new metaphors for understanding how apparent agency and intelligence might emerge from complex information processing systems without requiring actual consciousness or external entities. Modern AI can generate human-like responses and appear to possess knowledge and intentions while operating through purely mechanical processes.

Applied to entity encounters, this framework suggests that psychedelic substances might alter brain function in ways that create the appearance of external intelligence through enhanced pattern recognition, memory access, and creative synthesis. The entities encountered might represent sophisticated internal simulations generated by brain systems operating under altered neurochemical conditions.

Research on AI consciousness and the "hard problem" of machine consciousness reveals how difficult it is to determine whether apparently intelligent behavior reflects genuine consciousness or sophisticated but unconscious information processing. Similar challenges apply to evaluating whether psychedelic entities represent genuine consciousness or convincing simulations generated by altered brain function.[32]

However, this framework faces the challenge of explaining why entity encounters often feel more real and compelling than normal conscious experience rather than less so. If the entities represent degraded brain function, why do experiencers typically report enhanced rather than diminished sense of reality and meaning?

Anomaly proponents argue that if artificial systems can develop emergent properties that surprise their creators, biological consciousness under psychedelic influence might similarly access emergent capabilities that transcend normal cognitive limitations. The entities encountered might represent genuine emergent intelligences rather than mere simulations.

Mystical Traditions and Archetypal Psychology

The frameworks developed within mystical and psychological traditions provide rich contexts for understanding entity encounters as expressions of deep psychological and spiritual processes that operate independently of specific pharmacological triggers.

Carl Jung's analytical psychology includes detailed theories about archetypal figures that emerge from the collective unconscious during periods of psychological transformation. Jung's work with patients experiencing psychotic episodes and his own personal explorations of altered states revealed consistent patterns of encounter with autonomous psychological figures that seemed to possess independent agency and knowledge.[33]

Applied to psychedelic entity encounters, Jungian theory suggests that the entities might represent archetypal figures emerging from deep levels of the psyche that are normally inaccessible to conscious awareness. These figures could possess apparent autonomy and knowledge because they represent accumulated wisdom and intelligence that transcends individual conscious knowledge.

Religious and mystical traditions across cultures include sophisticated frameworks for understanding encounters with non-human intelligences during altered states of consciousness. Christian mysticism describes encounters with angels and saints, Buddhist meditation includes meetings with various categories of beings, and shamanic traditions worldwide involve contact with spirit guides and teachers.[34]

These traditional frameworks suggest that entity encounters represent normal rather than anomalous experiences when consciousness is freed from ordinary constraints through spiritual practices, meditation, or pharmacological means. The consistency of such reports across cultures and historical periods indicates either universal psychological processes or genuine contact with transpersonal domains of reality.

However, critics argue that traditional mystical frameworks, while psychologically meaningful, provide no scientific evidence for the actual existence of the entities described. The fact that similar experiences occur across cultures might reflect shared human neurology and psychology rather than contact with genuine external beings.

Systems Theory and Emergent Intelligence

Network science and systems theory offer frameworks for understanding how apparent intelligence and agency might emerge from complex interactions within distributed systems, potentially explaining entity encounters without requiring external consciousness or supernatural explanations.

Research on emergent properties in complex systems reveals that sophisticated behaviors and apparent intelligence can arise from interactions between simpler components following basic rules. Ant colonies, bird flocks, and neural networks all demonstrate forms of intelligence that exceed the capabilities of their individual components.[35]

Applied to psychedelic experiences, this framework suggests that entity encounters might represent emergent properties of brain networks operating under altered conditions. The apparent intelligence and autonomy of encountered entities might reflect sophisticated information processing capabilities that emerge when normal neural organization is disrupted and reorganized by psychedelic substances.

The model could explain why entities often seem to possess knowledge that exceeds the experiencer's normal capabilities—they might represent emergent intelligence arising from the integration of normally disconnected brain networks and memory systems. This enhanced integration could generate insights and problem-solving abilities that feel like communication with external sources.

However, this framework struggles to explain the apparent separateness and autonomy of encountered entities. If they represent emergent properties of the experiencer's own brain, why do they consistently appear as separate beings rather than enhanced aspects of the self? The phenomenology of otherness that characterizes entity encounters remains challenging for systems-based explanations.

Current State of Evidence: An Honest Assessment

After three decades of renewed scientific investigation into psychedelic entity encounters, what can be concluded about these extraordinary experiences? The picture that emerges is complex and contentious, with different researchers drawing dramatically different conclusions from similar evidence.

The phenomenon of entity encounters during psychedelic experiences is well-documented and remarkably consistent across different substances, users, and cultural contexts. Strassman's pioneering research established that approximately 50% of DMT users report such encounters, while studies of other psychedelics report lower but still significant frequencies of entity contact.

Contemporary neuroimaging research has revealed specific brain changes associated with psychedelic experiences, including alterations in connectivity between brain networks and changes in activity patterns in regions associated with self-awareness and social cognition. However, the relationship between these neural changes and subjective reports of entity encounters remains unclear.

The consistency of certain features across entity encounter reports—apparent autonomy, intelligence, communication abilities, and tendency to provide meaningful information—suggests either genuine contact with external intelligences or universal psychological processes that create similar experiences under similar neurochemical conditions.

Studies of the long-term effects of entity encounters reveal lasting changes in personality, worldview, and psychological well-being that often relate directly to guidance or information received from encountered entities. These effects suggest that the encounters, whatever their ultimate nature, engage fundamental aspects of human psychology and meaning-making.

However, no clear mechanism has been identified for how consciousness could contact genuinely external entities through psychedelic experiences. While various theoretical possibilities have been proposed—from quantum field effects to archetypal psychology—none have gained sufficient empirical support to provide convincing explanations for the alleged phenomena.

The challenge of distinguishing between genuine contact with external entities and sophisticated brain-generated simulations remains formidable. The subjective conviction that experiencers feel about the reality of their encounters provides compelling phenomenological evidence but cannot by itself establish the objective reality of the entities encountered.

The Information Transmission Question

One of the most intriguing aspects of psychedelic entity encounters involves reports of information transmission—cases where entities allegedly provide knowledge that was not previously accessible to the experiencer and that sometimes proves accurate or valuable upon subsequent investigation.

Ethnobotanist Dennis McKenna has documented cases where ayahuasca users report receiving information from plant spirits about medicinal properties of plants that they had not previously known. Some of these reports have allegedly led to the discovery of genuine therapeutic applications, though systematic verification of such claims remains limited.[36]

Strassman's research includes reports of entities providing scientific or technical information that subjects claim exceeded their normal knowledge. However, rigorous verification of such claims proves difficult because of the subjective nature of the information and the challenge of establishing that it was genuinely novel rather than creatively assembled from existing knowledge.

The information transmission question raises fundamental issues about the nature of knowledge, creativity, and the boundaries of individual consciousness. Even if the information comes from internal rather than external sources, the mechanisms by which psychedelics might enhance access to unconscious knowledge or creative problem-solving abilities deserve scientific investigation.

Critics argue that apparent information transmission might reflect enhanced pattern recognition, memory integration, or creative synthesis rather than contact with external sources. The conviction that information comes from external entities might result from the altered sense of self and agency that characterizes psychedelic experiences.

However, the consistency of reports across different users and the occasional apparent accuracy of transmitted information suggests that something more than random neural activity may be involved. Whether this represents enhanced internal processing or genuine external contact remains an open question requiring further investigation.

Therapeutic Implications and Clinical Applications

Regardless of their ultimate ontological status, psychedelic entity encounters have demonstrated significant therapeutic potential in clinical settings. Studies of psilocybin-assisted therapy for depression, PTSD, and addiction report that many patients describe receiving guidance or healing from entities encountered during their sessions.[37]

Research by Roland Griffiths at Johns Hopkins University has shown that mystical-type experiences, including entity encounters, correlate with positive therapeutic outcomes in psilocybin-assisted treatment for depression and anxiety. Patients who report meaningful contact with spiritual or transcendent beings during their sessions often show greater and more lasting improvement than those who do not have such experiences.[38]

The therapeutic value of entity encounters appears to operate independently of beliefs about their objective reality. Patients who interpret their experiences as genuine spiritual contact and those who understand them as meaningful psychological processes both report significant benefits from the guidance and insights received during their encounters.

This suggests that the healing potential of entity encounters may operate through psychological mechanisms related to meaning-making, self-reflection, and access to unconscious knowledge rather than through literal contact with external healing entities. However, the possibility that genuine transpersonal contact might contribute to healing effects cannot be entirely ruled out based on current evidence.

Clinical researchers emphasize the importance of set and setting in facilitating beneficial entity encounters while minimizing potentially disturbing or traumatic experiences. Proper preparation, supportive environments, and skilled therapeutic guidance appear crucial for maximizing therapeutic benefits while ensuring patient safety.

Future Directions: Questions for Further Investigation

Rather than settling questions about the reality of psychedelic entity encounters, current research has identified specific issues that require further investigation. If the encounters represent genuine contact with external intelligences, what mechanisms might enable such contact and how could they be scientifically studied? How could researchers distinguish between genuine entity contact and sophisticated brain-generated simulations?

If entity encounters result from altered brain function, how can researchers better understand the specific neural mechanisms involved and their relationship to the subjective experiences reported? What can these experiences reveal about the nature of consciousness, self-awareness, and social cognition under altered conditions?

The development of new research methods offers potential approaches to these questions. Advanced neuroimaging during psychedelic experiences might reveal specific neural correlates of entity encounters that could help distinguish between different explanatory models. Real-time monitoring of brain activity during reported entity communication could provide insights into the mechanisms involved.

Systematic studies of information transmission during entity encounters could help evaluate claims about receiving novel knowledge from external sources. Controlled protocols for documenting and verifying allegedly transmitted information might help determine whether such reports reflect genuine external communication or enhanced internal processing.

Cross-cultural research could help distinguish universal features of entity encounters from culture-specific interpretations and expectations. Comparative studies of entity encounters across different substances, cultural contexts, and individual characteristics might reveal underlying patterns that could inform theoretical understanding.

Perhaps most importantly, the investigation of psychedelic entity encounters highlights the need for expanded theoretical frameworks that can integrate findings from neuroscience, psychology, anthropology, and consciousness studies. Whether or not the entities represent genuine external intelligences, the study of these experiences reveals important aspects of human consciousness that challenge conventional understanding of the boundaries of self and reality.

The case of Sarah and her first DMT experience continues to puzzle researchers precisely because it combines elements that resist easy categorization: subjective conviction about external contact, apparent communication with autonomous beings, and lasting psychological impact that suggests genuine meaning and significance. Whether this case represents evidence for contact with non-human intelligences or an illustration of the extraordinary creative and meaning-making capabilities of human consciousness under altered conditions remains an open question—one that bears on fundamental issues about the nature of consciousness, reality, and the possibility of genuine transcendence.

Cultural Lens: From Plant Teachers to Machine Elves

The cross-cultural interpretation of psychedelic entity encounters reveals fundamental differences in how various societies understand consciousness, reality, and the possibility of contact with non-human intelligences. These cultural frameworks profoundly shape both the reported characteristics of encountered entities and their interpretation within different knowledge systems.

Indigenous Amazonian cultures view entity encounters as normal aspects of spiritual and healing practices rather than anomalous experiences requiring special explanation. The Shipibo-Conibo people describe detailed relationships with plant spirits that provide ongoing guidance about medicine, ecology, and social organization. These entities are understood as genuine inhabitants of spiritual realms rather than hallucinations or psychological projections.

In contrast, Western secular culture typically interprets entity encounters as products of altered brain chemistry, interesting perhaps for their psychological significance but lacking objective reality. This interpretive difference shapes not only how the experiences are understood but also how they are integrated into daily life and decision-making processes.

The terminology used to describe entities also reflects cultural differences in interpretation. Indigenous traditions emphasize relationships with "plant teachers," "animal spirits," and "ancestral guides" that connect to established spiritual cosmologies. Western psychedelic culture often describes "machine elves," "aliens," and "interdimensional beings" that reflect contemporary technological and science fiction frameworks.

These interpretive differences raise important questions about the relationship between cultural expectation and reported experience. Do cultural frameworks actually influence the types of entities encountered, or do they simply provide different language for describing similar underlying experiences? How might Western scientific assumptions about the nature of consciousness and reality limit the investigation of phenomena that transcend these frameworks?

The challenge for cross-cultural research is developing methodologies that can investigate entity encounters without imposing particular cultural assumptions about their nature or significance. This requires recognizing that different cultures may have developed valid knowledge systems for understanding aspects of consciousness that Western science has yet to explore adequately.

Chapter 12: Savant Abilities and Sudden Genius

On October 7, 1994, Dr. Tony Cicoria, a 42-year-old orthopedic surgeon, was making a phone call from a pay phone outside his family's lakeside cabin in Albany, New York, when lightning struck the building and traveled through the phone line. The electrical surge threw him backward, stopping his heart briefly before a nearby nurse successfully resuscitated him. Cicoria appeared to recover completely from what should have been a fatal encounter, returning to his surgical practice within days.[1]

But three weeks later, something extraordinary began to happen. Cicoria developed an overwhelming, almost obsessive desire to listen to piano music—particularly classical works he had never shown interest in before. Soon he was compulsing himself to play, despite having only rudimentary piano skills from childhood lessons decades earlier. Within months, he was composing complex original pieces, working for hours each day with an intensity that surprised his family and colleagues. Most remarkably, the music seemed to come to him fully formed, as if downloaded directly into his consciousness.[2]

This case, documented by neurologist Oliver Sacks, represents one of the most dramatic examples of acquired savant syndrome—the sudden emergence of extraordinary abilities following brain injury or trauma. While congenital savant abilities have been recognized for over a century, cases like Cicoria's challenge fundamental assumptions about the nature of human potential, learning, and the relationship between brain injury and cognitive enhancement.

Defining Savant Abilities and Sudden Genius

Savant syndrome, first described in detail by British physician John Langdon Down in 1887, refers to the occurrence of extraordinary abilities in specific domains alongside significant cognitive or developmental limitations. The condition was originally termed "idiot savant" before evolving to the more respectful "savant syndrome" to acknowledge both the remarkable capabilities and the dignity of individuals who possess them.[3]

Contemporary research distinguishes between several categories of savant abilities. Congenital savant syndrome appears in approximately 1 in 10 individuals with autism spectrum disorders and is characterized by exceptional skills that manifest early in development, often without explicit training or instruction. These abilities typically occur in specific domains including calendar calculation, rapid mathematical computation, hyperrealistic artistic reproduction, perfect pitch and musical improvisation, and extraordinary memory for specific categories of information.[4]

Acquired savant syndrome represents a much rarer phenomenon where exceptional abilities emerge suddenly following brain injury, illness, or other neurological events. Fewer than 50 well-documented cases exist in the scientific literature, making this one of the most unusual conditions in neuroscience. The abilities that emerge often mirror those seen in congenital savant syndrome but appear dramatically and without warning in previously normal individuals.[5]

Sudden genius represents a broader category that includes not only acquired savant abilities but also cases where individuals develop extraordinary creative or intellectual capabilities without obvious neurological triggers. These cases often involve artistic, musical, or literary breakthroughs that seem to exceed the individual's previous training or experience.

The defining characteristics of savant abilities include several distinctive features that distinguish them from ordinary talents or learned skills. The abilities often emerge without conscious effort or traditional learning processes, appearing almost instantly at expert levels of proficiency. They typically demonstrate remarkable precision and consistency, with performance that often exceeds that of trained professionals. The abilities frequently show obsessive or compulsive qualities, with individuals driven to exercise their newfound skills for hours each day.[6]

Perhaps most remarkably, savant abilities often coexist with significant limitations in other cognitive domains, creating what researchers call "islands of genius" surrounded by oceans of intellectual disability or dysfunction. This dissociation between extraordinary ability and general cognitive function challenges traditional assumptions about intelligence and expertise.

Cross-Cultural Patterns and Historical Perspectives

Reports of extraordinary abilities emerging suddenly or existing alongside developmental differences appear across cultures and throughout history, though their interpretation and social reception vary dramatically depending on cultural frameworks and beliefs about intelligence, disability, and human potential.

In many traditional societies, individuals with exceptional abilities coupled with social or cognitive differences are often viewed through spiritual or mystical lenses. Some Native American traditions recognize "spirit-touched" individuals who possess extraordinary gifts alongside unusual behaviors or social difficulties. These individuals may be seen as having special connections to spiritual realms that both gift them with remarkable abilities and create challenges in ordinary social functioning.[7]

Chinese culture has historically recognized the phenomenon of "奇才" (qicai) or "strange talent"—individuals whose extraordinary abilities in specific domains coexist with unusual personalities or social difficulties. Traditional Chinese medicine includes frameworks for understanding how exceptional mental abilities might emerge from imbalances or unusual configurations of vital energy, providing cultural contexts for accepting rather than pathologizing such differences.[8]

In contrast, Western medical tradition has typically focused on the disability aspects of savant syndrome, viewing the extraordinary abilities as curious anomalies rather than valuable gifts. This perspective is gradually shifting as research reveals the

242

sophisticated cognitive mechanisms underlying savant abilities and their potential applications for understanding human intelligence more broadly.

Indian classical traditions include concepts of "विशेष बुद्धि" (vishesha buddhi) or "special intelligence" that can manifest suddenly through spiritual practices, divine grace, or unusual life experiences. These frameworks provide cultural contexts for understanding sudden genius as natural rather than pathological phenomena, though they often emphasize spiritual rather than neurological explanations.

African traditional healing systems often recognize individuals with exceptional abilities as having received special gifts from ancestors or spirits. In some cultures, the emergence of extraordinary skills following illness or trauma is interpreted as evidence of spiritual calling or initiation into healing or divination practices.

However, cultural attitudes toward sudden genius and savant abilities are not uniformly positive. Some societies view such dramatic changes with suspicion or fear, interpreting them as evidence of spiritual possession, mental illness, or dangerous disruption of natural order. These negative interpretations can lead to social isolation or even persecution of individuals who develop extraordinary abilities.

The Scientific Investigation: From Curiosity to Understanding

The systematic study of savant abilities began in the late 19th century with case reports by physicians like John Langdon Down and later Alfred Tredgold, who documented individuals with extraordinary memory, calculation, or artistic abilities coexisting with intellectual disabilities. Early research focused primarily on description and classification rather than understanding underlying mechanisms.

The modern era of savant research began with the work of Bernard Rimland in the 1960s and has been advanced significantly by Darold Treffert, who has documented and studied savant abilities for over four decades. Treffert's research has identified consistent patterns across cases and proposed neurological mechanisms that might explain how extraordinary abilities can emerge alongside cognitive limitations.[9]

Contemporary neuroscience has brought sophisticated tools to bear on understanding savant abilities. Brain imaging studies using MRI, PET scans, and other techniques have revealed structural and functional differences in the brains of individuals with savant abilities compared to both neurotypical individuals and those with autism or other developmental conditions without savant skills.[10]

Research by Allan Snyder at the University of Sydney has used transcranial magnetic stimulation (TMS) to temporarily induce savant-like abilities in neurotypical individuals, providing experimental evidence for theories about the neural mechanisms underlying these extraordinary capabilities. Snyder's work suggests that savant abilities might result from accessing normally filtered sensory and cognitive information rather than developing entirely new capabilities.[11]

Studies of acquired savant syndrome have been particularly illuminating because they allow researchers to study the same individual before and after the emergence of extraordinary abilities. These cases provide unique windows into how brain changes can unlock human potential that was previously inaccessible.

The Scientific Debate: Three Distinct Explanatory Frameworks

The investigation of savant abilities and sudden genius has generated distinct positions within neuroscience and psychology, each offering different explanations for how extraordinary abilities can emerge and what they reveal about human potential.

Neuroplasticity Realists: Rewiring and Disinhibition

The mainstream neuroscientific position, represented by researchers like Alvaro Pascual-Leone, Allan Snyder, and most contemporary cognitive neuroscientists, argues that savant abilities result from well-understood brain mechanisms involving neural plasticity, compensation, and the disinhibition of normally suppressed capabilities.

Allan Snyder, professor of neuroscience at the University of Sydney, has developed the most comprehensive neurological theory of savant abilities. Snyder argues that the extraordinary skills result from "privileged access" to normally filtered sensory and cognitive information. In typical brain function, higher-level cognitive processes filter raw sensory data to focus attention on behaviorally relevant information. Damage to these filtering mechanisms, or their developmental absence, can allow access to much more detailed information than is normally available to consciousness.[12]

Snyder's research using transcranial magnetic stimulation provides experimental support for this theory. By temporarily suppressing activity in the left anterior temporal lobe—a region involved in conceptual processing and categorical thinking—Snyder's team has induced temporary savant-like abilities in neurotypical volunteers. These include enhanced drawing skills, improved proofreading abilities, and increased sensitivity to numerical patterns.[13]

Darold Treffert, while acknowledging the complexity of savant abilities, emphasizes neuroplasticity and compensation mechanisms in his explanations. Treffert argues that brain injury or developmental differences can cause neural circuits to reorganize in ways that create extraordinary capabilities in specific domains while impairing function in others. This reorganization may recruit additional brain areas for particular tasks, leading to enhanced performance that exceeds normal abilities.[14]

When neuroplasticity realists examine cases like Tony Cicoria, they emphasize that lightning strikes can cause subtle but significant brain damage that alters neural connectivity and function. The obsessive musical behavior and enhanced abilities may result from damage to inhibitory circuits combined with strengthening of neural pathways involved in musical processing and memory.

Pascual-Leone's research on brain stimulation and plasticity suggests that all human brains contain latent capabilities that are normally suppressed or inaccessible. Savant abilities may represent the expression of these latent capabilities when normal inhibitory mechanisms are compromised through injury or developmental differences.[15]

Neuroplasticity realists also point out that savant abilities typically come with significant trade-offs in other cognitive domains. The same brain changes that enable extraordinary skills often impair social functioning, executive control, or general intelligence, suggesting that the brain operates within resource constraints that make simultaneous optimization across all domains impossible.

Developmental Optimists: Hidden Potential and Enhancement

A second group of researchers, while accepting conventional neurological explanations, argues that savant abilities reveal untapped potential in human cognition that could potentially be accessed through training, technology, or other interventions without requiring brain damage or developmental disability.

Berit Brogaard, professor of philosophy and neuroscience at the University of Miami, studies acquired savant syndrome and argues that these cases demonstrate that the human brain contains far more potential than is typically expressed. Brogaard's research suggests that the abilities revealed in acquired savant syndrome were present but latent before the triggering event, indicating that human cognitive capacity may be significantly underestimated.[16]

Brogaard emphasizes cases where acquired savants demonstrate abilities that seem to exceed what could be explained by simply removing inhibitory constraints. Some individuals develop not just enhanced versions of normal abilities but entirely new forms of perception or cognition that suggest access to previously unknown capabilities.

Snyder himself has become increasingly optimistic about the possibility of safely inducing savant-like abilities through brain stimulation techniques. His research suggests that temporary enhancement of specific cognitive abilities may be possible without the trade-offs typically seen in natural savant syndrome. This could have profound implications for education, rehabilitation, and human enhancement.[17]

Teresa Schubert and Torkel Klingberg's research on cognitive training and brain plasticity suggests that intensive practice and appropriate training methods can dramatically enhance specific cognitive abilities in neurotypical individuals. While these improvements don't typically reach savant levels, they demonstrate that human cognitive potential is more malleable than traditionally assumed.[18]

When developmental optimists examine sudden genius cases, they emphasize the possibility that similar breakthroughs might be achievable through deliberate practice, cognitive training, or brain stimulation techniques. The obsessive focus and intense practice often seen in savant syndrome might be replicable through motivation and appropriate training protocols.

However, developmental optimists must address the consistent finding that natural savant abilities almost always coexist with significant cognitive or social limitations. If the abilities simply represent unleashed potential, why don't they appear without these accompanying difficulties? The trade-off pattern suggests that extraordinary abilities may require neural configurations that necessarily impair other functions.

Anomaly Theorists: Information Fields and Extended Mind

A third, more speculative group suggests that some savant abilities, particularly those involving access to complex information without apparent learning, may involve mechanisms that transcend conventional neuroscience and point toward extended or non-local models of consciousness and cognition.

Rupert Sheldrake's morphic field theory provides one framework for understanding how individuals might access complex information without prior learning. Sheldrake proposes that skilled performance in any domain creates morphic fields that can be accessed by other individuals with appropriate neural configurations. Savant abilities might represent enhanced sensitivity to these information fields rather than purely internal brain processes.[19]

While morphic field theory lacks mainstream scientific acceptance, some researchers point to aspects of savant abilities that are difficult to explain through conventional neural mechanisms alone. These include cases where savants demonstrate knowledge or skills that seem to exceed what could be learned through available experience or training.

Some cases of sudden musical genius, like Tony Cicoria's, involve the emergence of complex compositional abilities that typically require years of training and cultural immersion. The apparent full-formed nature of the musical ideas that emerged suggests possible access to information or creative processes that transcend individual brain function.

Dean Radin and other consciousness researchers have proposed that some savant abilities might involve quantum effects or non-local information processing that could enable access to information beyond what is normally available to individual brains. While highly speculative, these theories attempt to address cases where savant abilities seem to involve knowledge or skills that exceed what conventional learning could provide.[20]

However, anomaly theorists face significant challenges in providing specific mechanisms for how non-local information access might work or testable predictions that could distinguish their theories from conventional explanations. The tendency for savant abilities to appear in specific domains with clear neural correlates suggests that conventional brain-based explanations remain more parsimonious.

Frameworks from Part I: New Perspectives on Human Potential

The theoretical frameworks developed in Part I offer diverse approaches to understanding savant abilities and sudden genius, each providing different insights into the mechanisms and implications of these extraordinary phenomena.

The Predictive Brain: Filters, Precision, and Raw Data

The predictive processing model provides perhaps the most compelling contemporary framework for understanding savant abilities, particularly Allan Snyder's privileged access theory. If the brain normally operates by filtering sensory input through predictive models that emphasize behaviorally relevant information, savant abilities might result from reduced filtering that allows access to much more detailed raw data.

Research by Jakob Hohwy and Andy Clark suggests that consciousness emerges from the brain's predictive models of both external environments and internal states. In typical cognition, high-level predictions suppress detailed sensory information to focus attention on unexpected or behaviorally relevant signals. Savant abilities might result from disrupted predictive hierarchies that allow normally suppressed information to reach consciousness.[21]

Applied to specific savant abilities, this framework explains how calendar calculators might access detailed temporal patterns that are normally filtered from awareness, how artistic savants might perceive visual details that are typically ignored, and how musical savants might detect acoustic patterns that exceed normal conscious access.

The predictive processing model also explains why savant abilities often involve repetitive, stereotyped behaviors. If predictive filtering is reduced, the resulting flood of sensory information might be managed through rigid behavioral routines that provide structure and predictability in an otherwise overwhelming perceptual environment.

However, the framework struggles to explain cases where savant abilities involve creative or generative capabilities rather than just enhanced perception or memory. Tony Cicoria's musical compositions suggest active creation rather than passive access to existing information, requiring additional mechanisms beyond simple filter removal.

Critics also note that predictive processing models don't obviously explain why filter disruption would enhance abilities in specific domains rather than creating general cognitive chaos. The selective nature of savant abilities suggests more targeted mechanisms than simple global disinhibition.

Consciousness Models: Distributed Intelligence and Access

Theories about the fundamental nature of consciousness raise questions about the boundaries of individual intelligence and the possibility that savant abilities represent access to forms of information or processing that transcend individual brains.

Global Workspace Theory, developed by Bernard Baars and Stanislas Dehaene, suggests that consciousness emerges from the global broadcasting of information across multiple brain networks. Savant abilities might result from altered global workspace function that provides unusual access to specialized processing modules that are normally unconscious.[22]

This framework could explain how savants access complex calculations, detailed memories, or sophisticated pattern recognition capabilities that seem to exceed conscious cognitive capacity. The abilities might represent direct conscious access to unconscious processing systems that normally operate below the threshold of awareness.

Integrated Information Theory's emphasis on information integration across multiple brain systems provides another perspective on savant abilities. If consciousness involves the integration of information from diverse processing systems, altered integration patterns might create access to normally unavailable cognitive capabilities.[23]

However, consciousness-based theories must explain why altered access patterns would enhance rather than disrupt cognitive function. If savant abilities represent unusual consciousness configurations, why do they typically produce superior rather than merely different performance in specific domains?

The theories also struggle with the trade-off patterns consistently observed in savant syndrome. If the abilities simply represent alternative consciousness configurations, why do they almost always coexist with significant limitations in other cognitive domains?

Quantum Mechanics and Coherent Processing

Some researchers have proposed quantum effects as potential mechanisms for savant abilities, particularly those involving rapid calculation, pattern recognition, or access to complex information without apparent learning processes.

Stuart Hameroff and Roger Penrose's Orchestrated Objective Reduction theory suggests that consciousness emerges from quantum processes in neural microtubules. Altered quantum processing might enable the kind of rapid, massively parallel computation that characterizes some savant abilities, particularly calendar calculation and mathematical skills.[24]

Henry Stapp's quantum theories of consciousness propose that conscious observation involves quantum measurement processes that could theoretically enable access to information through non-classical pathways. Savant abilities might represent enhanced quantum information processing that exceeds normal cognitive limitations.[25]

However, most quantum physicists remain skeptical of macroscopic quantum effects in biological systems. Max Tegmark's calculations suggest that quantum coherence in neural structures would decohere far too rapidly to support sustained quantum computation of the type that would be required for savant abilities.[26]

Even if quantum effects operate in the brain, they would need to explain why savant abilities cluster in specific domains rather than enhancing general cognitive function. The selectivity of savant abilities suggests more targeted mechanisms than general quantum enhancement would provide.

Artificial Intelligence and Specialized Processing

The development of AI systems with extraordinary capabilities in specific domains while lacking general intelligence provides useful analogies for understanding savant abilities. Modern AI demonstrates how specialized processing systems can achieve superhuman performance in narrow domains without general intelligence or consciousness.

Deep learning systems like AlphaGo achieve extraordinary performance in specific domains through massive parallel processing focused on narrow problem spaces. Savant abilities might involve similar specialization where neural resources are concentrated on particular types of information processing at the expense of general cognitive flexibility.[27]

The AI analogy suggests that savant abilities might represent extreme examples of cognitive specialization rather than access to entirely new types of information or processing. Just as AI systems achieve superhuman performance through focused optimization, savant brains might achieve extraordinary abilities through neural configurations that sacrifice breadth for depth.

However, the AI analogy struggles to explain the sudden emergence of abilities in acquired savant syndrome. AI systems require extensive training to achieve specialized performance, while acquired savants often demonstrate abilities immediately following brain injury or trauma.

The analogy also doesn't address the apparently effortless nature of many savant abilities. AI systems require massive computational resources to achieve superhuman performance, while savants often demonstrate extraordinary abilities with apparent ease and minimal conscious effort.

Network Dynamics and Reorganization

Systems theory and network science provide frameworks for understanding how brain injury or developmental differences might lead to neural reorganization that creates extraordinary capabilities through altered connectivity patterns and information flow.

Research by Steven Strogatz and others on network dynamics reveals how complex systems can undergo phase transitions that dramatically alter their functional properties. Brain injury might trigger similar phase transitions that reorganize neural networks in ways that create new functional capabilities.[28]

The small-world network properties of brain connectivity suggest that relatively minor changes in connection patterns could have dramatic effects on information processing capabilities. Savant abilities might result from reorganized connectivity that creates more efficient processing pathways for specific types of information.

Olaf Sporns' research on brain network topology suggests that different connectivity patterns are optimized for different types of cognitive functions. Savant abilities might represent network configurations that are highly optimized for specific cognitive domains at the expense of general-purpose processing.[29]

However, network-based explanations must account for why reorganization typically enhances rather than merely alters cognitive function. If savant abilities simply reflect different network configurations, why do they represent superior rather than just alternative performance patterns?

The framework also needs to explain the speed with which abilities can emerge in acquired savant syndrome. Network reorganization typically requires time and experience, while some acquired savant abilities appear almost immediately following triggering events.

Current State of Evidence: What We Know and Don't Know

After more than a century of research into savant abilities and sudden genius, what can be concluded about these extraordinary phenomena? The picture that emerges reveals both remarkable consistencies and persistent mysteries that continue to challenge our understanding of human potential and brain function.

The basic phenomenology of savant abilities is well-established and remarkably consistent across cases. The abilities cluster in specific domains including calendar calculation, mathematical computation, musical performance, artistic reproduction, and exceptional memory. They often emerge without explicit training and demonstrate levels of performance that exceed those of trained professionals.

Contemporary neuroscience has identified several brain mechanisms that contribute to savant abilities. Reduced left-hemisphere function and increased right-hemisphere activation appear in many cases, along with altered connectivity between brain regions and changes in inhibitory processing. These findings support theories emphasizing neural reorganization and disinhibition as primary mechanisms.

The trade-off pattern between extraordinary abilities and general cognitive function appears consistently across cases, suggesting that savant abilities emerge from neural configurations that necessarily limit other cognitive functions. This pattern challenges simple enhancement theories and suggests fundamental constraints on cognitive optimization.

However, significant mysteries remain. The speed with which abilities can emerge in acquired savant syndrome suggests mechanisms that operate more rapidly than typical learning or neural plasticity. The effortless nature of many savant abilities contrasts with the intensive effort typically required to develop expertise through normal training.

Perhaps most puzzling are cases where savant abilities seem to involve access to information or skills that exceed what could reasonably be acquired through available experience. While such cases remain controversial and may be explained through subtle learning processes, they suggest possibilities that current neuroscience cannot fully address.

The relationship between obsessive behavior and extraordinary ability in savant syndrome also remains incompletely understood. While intense focus clearly contributes to skill development, the compulsive nature of savant interests suggests mechanisms beyond simple motivation or attention.

The Enhancement Question: Implications for Human Potential

The study of savant abilities raises profound questions about human potential and the possibility of safely enhancing cognitive capabilities without the disabilities that typically accompany natural savant syndrome. If the abilities represent unleashed potential rather than entirely new capabilities, they might be accessible through training, technology, or other interventions.

Allan Snyder's success in temporarily inducing savant-like abilities through transcranial magnetic stimulation provides proof-of-concept evidence that at least some aspects of savant abilities can be accessed in neurotypical individuals. This research suggests that human cognitive potential may be significantly constrained by neural inhibition that could potentially be modulated.[30]

However, the consistent trade-off patterns in natural savant syndrome raise questions about whether extraordinary abilities can be achieved without sacrificing other cognitive functions. The brain appears to operate within resource constraints that make simultaneous optimization across all domains challenging or impossible.

Research on cognitive training and brain plasticity suggests that intensive practice can enhance specific abilities beyond normal levels, though typically not to savant levels. The combination of training with brain stimulation or other technologies might potentially achieve greater enhancement than either approach alone.

The ethical implications of cognitive enhancement research based on savant abilities are complex. While the potential benefits of enhancing human capabilities are significant, the association of these abilities with disability raises questions about the value and acceptability of different forms of human diversity.

Cross-Cultural Perspectives on Genius and Abnormality

The interpretation and social reception of savant abilities vary dramatically across cultures, reflecting different beliefs about intelligence, disability, and human potential. These cultural differences have important implications for how individuals with such abilities are treated and how their capabilities are understood and utilized.

Western medical culture has traditionally emphasized the disability aspects of savant syndrome, viewing the extraordinary abilities as curious anomalies accompanying more significant impairments. This perspective is gradually changing as research reveals the sophisticated mechanisms underlying savant abilities and their potential applications.

Many non-Western cultures provide more positive frameworks for understanding individuals with exceptional abilities alongside social or cognitive differences. Some traditions view such individuals as possessing special gifts or spiritual connections that should be honored rather than pathologized.

Indigenous healing traditions often recognize individuals with unusual abilities as having special roles in community spiritual and healing practices. The emergence of extraordinary capabilities following illness or trauma may be interpreted as evidence of spiritual calling rather than brain damage.

However, cultural attitudes are not uniformly positive. Some societies view sudden changes in ability or behavior with suspicion, interpreting them as evidence of spiritual possession, mental illness, or dangerous disruption of natural order.

These cultural variations highlight the importance of social context in determining whether exceptional abilities are viewed as gifts to be cultivated or problems to be managed. The same neurological condition may lead to very different life outcomes depending on cultural frameworks and social support systems.

Future Directions: Questions for Further Investigation

Rather than settling questions about the nature and mechanisms of savant abilities, current research has identified specific issues that require further investigation. How can the speed of ability emergence in acquired savant syndrome be explained through conventional neural mechanisms? What accounts for the apparent effortlessness of savant performance compared to normal expertise development?

If savant abilities represent access to normally suppressed information or capabilities, how can these capabilities be safely accessed without the disabilities that typically accompany them? What are the fundamental constraints that create trade-offs between extraordinary abilities and general cognitive function?

The development of new research methods offers potential approaches to these questions. Advanced neuroimaging during skill performance might reveal the specific neural mechanisms that enable savant abilities. Real-time monitoring of brain activity during ability emergence could provide insights into the speed and nature of neural changes.

Studies combining brain stimulation with intensive training might help determine whether savant-like abilities can be induced safely and sustainably in neurotypical individuals. Cross-cultural research could help distinguish universal features of savant abilities from culture-specific interpretations and expectations.

Perhaps most importantly, the investigation of savant abilities highlights the need for expanded theoretical frameworks that can integrate findings from neuroscience, cognitive psychology, and consciousness studies. Whether or not the abilities involve anomalous mechanisms, they reveal important aspects of human potential that challenge conventional understanding of intelligence, learning, and the boundaries of cognitive capability.

The case of Tony Cicoria continues to puzzle researchers precisely because it combines elements that resist easy explanation: the speed of ability emergence, the sophisticated nature of the musical abilities, and the apparent full-formed quality of the creative output. Whether this case represents the expression of latent potential, the result of specific neural reorganization, or access to information through mechanisms not yet understood by science remains an open question—one that bears on fundamental issues about human potential and the nature of genius itself.

Cultural Lens: When Genius Meets Disability

The cross-cultural interpretation of savant abilities reveals fundamental differences in how various societies understand intelligence, disability, and human value. These cultural frameworks profoundly shape both the life experiences of individuals with savant abilities and the scientific investigation of these phenomena.

Western biomedical culture typically emphasizes diagnostic categories and deficit-based models that focus on what individuals with savant syndrome cannot do rather than celebrating their extraordinary capabilities. This framework shapes both research priorities and social services, often leading to interventions designed to normalize behavior rather than cultivate exceptional abilities.

In contrast, many indigenous cultures recognize individuals with unusual combinations of abilities and limitations as having special spiritual or social roles. The Inuit concept of "angakkuit" includes individuals with extraordinary perceptual or cognitive abilities who may also have social or communication difficulties, but who are valued for their unique contributions to community life.

Hindu traditions include concepts of "विशेष शक्ति" (vishesha shakti) or "special powers" that can emerge through spiritual practice, divine grace, or unusual life experiences. These frameworks provide cultural contexts for understanding extraordinary abilities as natural expressions of human potential rather than medical anomalies.

The terminology used to describe these conditions also reflects cultural values and assumptions. The evolution from "idiot savant" to "savant syndrome" in Western medicine reflects growing recognition of the dignity and value of individuals with these conditions, while many non-Western languages lack equivalent deficit-focused terminology.

These cultural differences raise important questions about how social frameworks influence both the development and expression of exceptional abilities. Do cultures that celebrate cognitive diversity actually facilitate the emergence and development of savant abilities? How might deficit-focused medical models inadvertently limit the potential of individuals with these extraordinary capabilities?

The challenge for cross-cultural research is developing frameworks that can study savant abilities without imposing particular cultural assumptions about intelligence, disability, or human value. This requires recognizing that different cultures may have developed valuable insights about human potential that complement rather than conflict with scientific understanding.

Part III

Toward a New Inquiry

Living with uncertainty as a path to deeper understanding. The investigation of anomalous phenomena reveals both the power and limitations of current scientific methods. This final section explores how to maintain intellectual humility while pursuing rigorous inquiry, how to synthesize insights across disciplines through "Venn Thinking," and how to cultivate wonder as a driving force for discovery. Rather than reaching definitive conclusions, we end with questions—and with frameworks for continuing the exploration of mysteries that may transform our understanding of mind, consciousness, and the nature of reality itself.

Chapter 13: Living with Mystery

In 2018, Ukrainian speleologists mapped the Veryovkina Cave to a depth of 2,212 meters—the deepest known cave on Earth. Even with modern technology, it took years of planning and dozens of expeditions to chart the twisting, dangerous passages. The team discovered entire sections they hadn't expected: underground lakes, unique microclimates, and life forms adapted to total darkness. What struck the explorers most was not what they found, but the realization that each mapped chamber revealed new passages leading deeper into the unknown.

Our exploration of reality follows a remarkably similar pattern. We illuminate some passages of understanding, map them in detail with careful observation and rigorous method, and yet consistently find new corridors we didn't know existed. Each discovery reveals the vastness of what remains unseen—and the profound humility needed to keep going.

This humility is not a retreat from knowledge but its highest expression. Consider the Hubble Deep Field observations, where astronomers pointed their most powerful telescope at what appeared to be empty space for eleven days in 1995. That seemingly vacant patch of sky revealed over three thousand galaxies, each containing billions of stars. The observation fundamentally changed our understanding of cosmic density and the universe's age. What we thought was emptiness was actually teeming with previously undetectable complexity.

Similarly, the Human Genome Project, completed in 2003, was supposed to provide a comprehensive blueprint of human biology. Instead, it revealed that humans have far fewer genes than expected and that vast regions of "junk DNA" perform crucial regulatory functions we're still discovering. Each genetic revelation opens new questions about epigenetic inheritance, gene expression networks, and the relationship between genotype and phenotype.

These examples share a crucial characteristic with the phenomena we've explored throughout this book: initial findings consistently reveal deeper layers of complexity rather than simple, definitive answers. This pattern suggests something fundamental about the nature of inquiry itself.

The Value of Uncertainty

Modern science is built on what philosopher Karl Popper called "provisional knowledge"—every theory is a model, not a final truth. The best researchers understand their work as waypoints in an ongoing journey, not destinations. Yet in popular culture, we often expect science to provide definitive answers, especially when it touches on extraordinary claims or challenges our fundamental assumptions about reality.

Living with mystery means resisting the twin temptations that bracket genuine inquiry: the skeptical impulse to dismiss anomalies prematurely and the credulous tendency to accept extraordinary claims without sufficient evidence. It means adopting what cognitive scientist John Dewey called "the discipline of suspended judgment"—a willingness to sit with uncertainty while continuing to investigate.

This suspension is not passive waiting but active engagement with complexity. When Thomas Kuhn examined scientific revolutions in *The Structure of Scientific Revolutions*, he observed that paradigm shifts often begin with anomalies that don't fit existing frameworks. The most significant advances come not from defending established theories but from taking anomalies seriously enough to investigate them rigorously.

Consider how quantum mechanics emerged from puzzling observations that classical physics couldn't explain—blackbody radiation curves, the photoelectric effect, atomic stability. Rather than dismissing these anomalies, physicists like Max Planck and Albert Einstein took them seriously, leading to theoretical frameworks that fundamentally transformed our understanding of reality. The lesson is not that every anomaly leads to revolution, but that maintaining openness to genuine surprises is essential for scientific progress.

What Our Theoretical Toolkit Revealed

In Part I, we assembled six conceptual frameworks that serve as tools for approaching the mysterious:

Predictive processing showed us how the brain constantly generates and updates models of the world, actively constructing rather than passively receiving experience. This framework illuminated how both ordinary perception and extraordinary experiences might emerge from the same fundamental cognitive architecture. When we examined near-death experiences through this lens, we saw how dying brains might generate vivid, meaningful experiences through predictive mechanisms responding to unprecedented neural conditions.

Theories of consciousness reminded us that despite centuries of scientific progress, we still lack consensus on whether mind emerges from matter or represents a fundamental feature of the universe. The "hard problem" of consciousness—explaining why there is subjective experience at all—remains unsolved. This theoretical uncertainty means that phenomena involving consciousness, from mystical experiences to apparent telepathy, cannot be dismissed simply because they challenge materialist assumptions.

Quantum mechanics and emergence theory illustrated how complexity, observation, and measurement can shape reality in ways we don't fully grasp. The quantum measurement problem—why quantum superpositions collapse into definite states when observed—suggests that the relationship between consciousness and physical reality may be more intricate than classical physics assumed. While quantum theories of mind remain speculative, they provide conceptual space for phenomena that seem impossible under strictly classical assumptions.

Artificial intelligence analogies provided mirrors for understanding our own cognition while raising fundamental questions about selfhood, agency, and the nature of intelligence. As AI systems become more sophisticated, they illuminate aspects of human consciousness we previously took for granted.

The distributed processing in neural networks, for instance, offers models for how collective intelligence might emerge from individual minds.

The science of subjective realities highlighted how mystical experiences, while culturally shaped, show consistent neurological patterns across traditions. Research on meditation, psychedelic states, and contemplative practices reveals that altered states of consciousness follow discoverable principles, even when their interpretation varies dramatically across cultures.

Systems and network theory gave us frameworks for understanding how interconnectedness—in brains, societies, and ecosystems—can generate unexpected, emergent phenomena. The behavior of complex systems often cannot be predicted from knowledge of individual components, suggesting that some apparently anomalous phenomena might reflect higher-order organizational principles we don't yet understand.

These tools don't settle the mysteries we've examined, but they provide language, models, and empirical approaches for engaging with them seriously. More importantly, they demonstrate that rigorous investigation can proceed even when fundamental questions remain unresolved.

What the Phenomena Taught Us

In Part II, we applied these theoretical tools to six categories of experiences that persist across cultures and resist simple explanation:

Near-death experiences, with their consistent features across diverse populations, challenge simple reductive explanations while remaining difficult to study under controlled conditions. The research of physicians like Sam Parnia and Pim van Lommel provides rigorous documentation of experiences during cardiac arrest, while critics like Susan Blackmore offer compelling neurobiological alternatives. What emerges from this debate is not certainty about what NDEs represent, but appreciation for the complexity of consciousness during extreme physiological states.

Reincarnation research, particularly the work of Ian Stevenson and Jim Tucker with children who report previous-life memories, presents cases that are difficult to explain through conventional mechanisms while remaining vulnerable to alternative explanations involving cryptomnesia, unconscious cueing, and cultural suggestion. The phenomenon illustrates how the same evidence can support radically different interpretations depending on one's theoretical commitments.

Dreams and precognition studies reveal how time perception, probability assessment, and memory construction interact in complex ways that challenge simple notions of linear causality. Dean Radin's meta-analyses suggest statistical anomalies in precognition experiments, while critics point to methodological issues and publication bias. The debate highlights fundamental questions about the nature of time and information processing.

Telepathy and collective mind phenomena touch on questions about the boundaries of individual consciousness and the possibility of direct mind-to-mind communication. Research on crisis telepathy and apparent psychic connections between identical twins suggests patterns that merit investigation, while critics emphasize the prevalence of coincidence and confirmation bias in human pattern recognition.

Psychedelic encounters with apparent non-human intelligences raise profound questions about the relationship between neurochemistry and subjective reality. The consistency of entity encounters across different individuals and substances suggests either universal features of altered consciousness or genuine contact with non-ordinary realities.

Savant abilities and acquired savantism demonstrate that human cognitive potential may be far more plastic and accessible than previously assumed, while challenging standard models of learning and development.

If these investigations share a common lesson, it is this: the frontier between known and unknown is not a fixed boundary but a dynamic zone where science, culture, and personal experience continually interact and reshape each other.

The Replication Crisis and Scientific Humility

Recent years have witnessed what researchers call the "replication crisis"—the discovery that many published findings in psychology, neuroscience, and even physics fail to hold up under repeated testing. In psychology, a large-scale replication project found that only about half of published studies could be successfully reproduced.[1] Similar challenges have emerged in neuroscience, where small sample sizes and complex statistical analyses can generate false positives.[2]

This crisis might seem to undermine confidence in scientific method. Instead, it demonstrates that science is working as intended. Self-correction through independent replication is built into the scientific process, even when it takes time to operate effectively. The crisis has led to important methodological reforms: larger sample sizes, pre-registered hypotheses, open data sharing, and more sophisticated statistical approaches.

But replication failures also remind us that our current methods are imperfect, especially when studying rare, complex, or deeply subjective phenomena. The experiences examined in this book often resist the controlled conditions that make replication straightforward. Near-death experiences cannot be induced ethically in laboratories. Spontaneous telepathic experiences don't occur on demand. Precognitive dreams cannot be scheduled for convenient observation.

This methodological challenge requires what anthropologist Clifford Geertz called "thick description"—detailed, contextualized accounts that capture the complexity of phenomena rather than reducing them to simple variables.[3] It also suggests the value of what philosopher of science Helen Longino terms "cognitive democracy"—incorporating multiple perspectives and knowledge traditions to develop more complete understanding.[4]

Indigenous knowledge systems, contemplative traditions, and experiential practices may offer insights that complement laboratory-based approaches. This doesn't mean abandoning scientific rigor, but rather expanding our methodological repertoire to match the complexity of what we're trying to understand.

Why Scientific Humility Is Strength

In public discourse, admitting uncertainty can appear as weakness. Politicians who change positions based on new evidence are criticized as "flip-floppers." Experts who acknowledge limitations in their knowledge may be dismissed as unreliable. This cultural bias toward certainty creates pressure to offer definitive answers even when evidence remains incomplete.

In science, however, uncertainty acknowledgment represents sophistication rather than weakness. Richard Feynman captured this principle in his famous observation: "I would rather have questions that can't be answered than answers that can't be questioned."[5] The ability to maintain productive uncertainty—what Feynman called "living with doubt"—enables continued discovery.

This intellectual virtue extends beyond professional scientists. In an interconnected world where complex problems require interdisciplinary collaboration, the capacity to hold multiple perspectives simultaneously becomes essential. Climate science, pandemic response, artificial intelligence development, and global economic policy all require integrating knowledge across disciplines while acknowledging significant uncertainties.

When we replace the demand for premature certainty with cultivated curiosity, we open ourselves to genuine discovery. We also develop resilience against misinformation, ideology, and the human tendency to cling to comforting but oversimplified answers.

Consider how the acceptance of continental drift required geologists to hold their theories lightly. Alfred Wegener's initial proposal was widely rejected because no mechanism for continental movement could be imagined. Only when plate tectonics provided that mechanism did the geological community embrace what evidence had long suggested. The lesson is not that every rejected theory will eventually be vindicated, but that maintaining theoretical flexibility enables paradigm shifts when evidence warrants them.

Philosophical Frameworks for Uncertainty

Contemporary philosophy of science offers several frameworks for understanding how knowledge develops under conditions of uncertainty. Karl Popper's concept of "clouds and clocks" distinguished between predictable, mechanistic systems (clocks) and inherently probabilistic, complex systems (clouds).[6] Many of the phenomena we've examined may belong to the cloud category, where precise prediction is impossible but pattern recognition remains valuable.

Thomas Kuhn's analysis of scientific revolutions showed how "normal science" operates within established paradigms until anomalies accumulate sufficiently to trigger paradigm shifts.[7] The phenomena examined in this book might represent such anomalies—observations that don't fit current paradigms but haven't yet generated compelling alternatives.

Philosopher Paul Feyerabend argued for "methodological pluralism," suggesting that scientific progress requires multiple, sometimes contradictory approaches operating simultaneously.[8] This perspective validates the coexistence of different research programs investigating the same phenomena through different theoretical lenses.

These philosophical frameworks suggest that uncertainty is not a temporary inconvenience to be eliminated but a permanent feature of inquiry into complex phenomena. The goal is not to eliminate uncertainty but to engage with it productively.

Revisiting Our Central Questions

As we near the end of our investigation, it's worth returning to some key questions that emerged throughout our journey, recognizing how each retains elements of unresolved mystery:

How does consciousness relate to the physical brain? Despite decades of neuroscientific progress, the hard problem of consciousness remains unsolved. We can map neural correlates of awareness, but the emergence of subjective experience from objective processes remains mysterious.

What is the nature of time and causality? Quantum mechanics suggests that our intuitive understanding of linear time may be incomplete. Retrocausality experiments and precognition research hint at temporal relationships we don't fully comprehend.

How do individual minds relate to collective intelligence? Network neuroscience reveals how individual brains connect to form social networks, but the mechanisms of collective cognition remain largely unexplored.

What are the limits of human cognitive potential? Savant abilities and acquired savantism suggest that ordinary consciousness may represent only a fraction of our cognitive capacity.

How do subjective experiences relate to objective reality? The consistency of mystical experiences across cultures suggests either universal features of consciousness or genuine contact with non-ordinary dimensions of reality.

Each of these questions has generated significant research and theoretical development, yet none admits simple answers. This persistence of mystery is not a failure of science but recognition of the profound depth of the phenomena we're investigating.

Contemporary Discoveries and Ongoing Mysteries

Recent scientific developments continue to reveal new layers of complexity rather than simple closure. The discovery of gravitational waves confirmed Einstein's predictions while opening entirely new domains of astronomical observation. CRISPR gene editing technology enables precise genetic modification while raising unprecedented questions about human enhancement and identity.

In neuroscience, optogenetics allows researchers to control individual neurons with light, yet consciousness remains as mysterious as ever. Brain organoids grown from stem cells develop neural networks resembling human brain structures, but we still don't understand how subjective experience emerges from neural activity.

These developments follow the pattern we've observed throughout this investigation: each answer generates new, often deeper questions. The most exciting frontiers are not where we're approaching final answers but where we're discovering new forms of complexity.

The Bridge to Synthesis

The chapters that follow will demonstrate how the most fruitful insights emerge when we stop asking "Which explanation is correct?" and start asking "What happens when we place multiple explanations in conversation with each other?" This approach—which we'll call **Venn Thinking**—doesn't collapse differences into a single narrative but uses their intersection to generate new possibilities.

Living with mystery, in this sense, isn't about resignation or intellectual surrender. It's about maintaining the delicate balance between healthy skepticism and genuine openness that enables discovery. It requires holding space for complexity, for the provisional nature of all knowledge, and for the possibility that the next passage in our conceptual cave might illuminate territories we never imagined existed.

The deepest caves on Earth continue to reveal new chambers to explorers willing to venture into unmapped darkness with proper equipment and careful method. Similarly, the deepest mysteries of consciousness and reality continue to reward investigators who combine rigorous analysis with genuine wonder. The adventure of discovery requires both tools and courage—and the wisdom to know that every illuminated passage reveals new depths waiting to be explored.

Chapter 14: Venn Thinking — A New Model for Knowing

In the late 16th century, cartographers faced a profound puzzle. Spanish, Portuguese, and Dutch explorers each returned with maps of the New World—all partially accurate, all partially wrong. Coastlines were distorted, islands misplaced, mountain ranges exaggerated or missing entirely. Each map reflected the explorer's specific route, available tools, cultural preconceptions, and the particular challenges they encountered.

No single map could be trusted in isolation. The Spanish maps excelled at documenting Pacific coastlines but poorly represented the Atlantic approaches. Portuguese charts captured excellent detail of Brazilian harbors but left vast inland territories blank. Dutch navigational maps provided precise coastal measurements but ignored indigenous knowledge of interior geography.

The breakthrough came when cartographers began overlaying these imperfect maps—tracing common coastlines, noting where two or three sources agreed, identifying systematic biases in each tradition. Where the Spanish map showed a bay and the Portuguese chart indicated a river mouth, careful examination often revealed a complex estuary system that neither had fully captured. The Americas took shape with far greater accuracy through synthesis than any single expedition could achieve.

The resulting maps were never perfect, but they were substantially better than any individual account. More importantly, the overlay process revealed exactly where knowledge remained incomplete, directing future expeditions toward the most productive areas of investigation.

This process embodies what we call **Venn Thinking**: the deliberate practice of overlapping perspectives, data sets, and conceptual frameworks to reveal patterns invisible to any single lens. It's not mere eclecticism—the uncritical mixing of different approaches—but a systematic method for extracting insights from the intersection of rigorous but incomplete knowledge systems.

Why We Need a New Model for Knowing

Our contemporary intellectual landscape is increasingly fractured by disciplinary specialization. Neuroscience rarely speaks the language of theology. Quantum physics maintains careful distance from psychology. Anthropology observes patterns that biology overlooks, while biology reveals mechanisms that anthropology doesn't investigate. Each field has developed sophisticated tools and vocabularies, but these very achievements can create barriers to cross-disciplinary insight.

When investigating anomalous phenomena—whether near-death experiences, sudden genius, or apparent telepathy—these disciplinary boundaries become particularly problematic. Proponents and skeptics alike can become trapped within their own explanatory frameworks, dismissing insights from adjacent fields without serious consideration. A neuroscientist might reject anthropological observations about universal patterns in mystical experience as "mere cultural construction." An anthropologist might dismiss neuroscientific findings about temporal lobe activity during religious experiences as "reductionist materialism."

This mutual dismissal impoverishes our understanding. The most complex phenomena—consciousness, creativity, healing, social coordination—operate at the intersection of multiple domains. They require what philosopher of science Helen Longino calls "cognitive democracy": the integration of different knowledge traditions and methodological approaches to develop more complete understanding.[1]

Venn Thinking provides a systematic approach to this integration. Rather than forcing different perspectives into artificial agreement, it asks: What do we learn when we place multiple rigorous frameworks in conversation with each other? How do their intersections illuminate aspects of phenomena that remain invisible to each approach in isolation?

Historical Precedents for Synthetic Insight

The most transformative scientific discoveries often emerge from the intersection of previously separate domains. The elucidation of DNA's structure required the convergence of chemistry, biology, physics, and mathematics. Rosalind Franklin's X-ray crystallography (physics) revealed the helical structure, while Erwin Chargaff's biochemical analysis (chemistry) identified base-pairing rules. James Watson and

Francis Crick's biological insight about complementary strands integrated these findings with Maurice Wilkins' physical measurements. No single discipline could have achieved this breakthrough alone.[2]

Similarly, the development of plate tectonics theory required integrating evidence from geology, oceanography, paleontology, and geophysics. Alfred Wegener's initial continental drift hypothesis was rejected largely because it remained confined to geological observation. Only when seafloor spreading provided a physical mechanism (geophysics), when magnetic striping patterns confirmed the timeline (paleomagnetism), and when fossil distributions supported ancient connections (paleontology) did the scientific community embrace the theory.[3]

More recently, the discovery of neuroplasticity emerged from the intersection of neuroscience, psychology, and rehabilitation medicine. Neuroscientists identified mechanisms of synaptic change, psychologists documented behavioral plasticity throughout the lifespan, and rehabilitation specialists observed functional recovery that seemed impossible under earlier brain models. Each perspective was necessary; none was sufficient.[4]

These cases demonstrate that synthetic thinking is not a luxury but a necessity for addressing complex phenomena. They also reveal the key characteristics of productive interdisciplinary synthesis: respect for the rigor of each contributing field, attention to empirical convergence, and willingness to revise assumptions when evidence from multiple domains points in unexpected directions.

Distinguishing Venn Thinking from Superficial Pluralism

Venn Thinking must be distinguished from what sociologist Mary Douglas calls "grid-group thinking"—the tendency to merge different frameworks without careful attention to their internal logic or empirical foundations.[5] True synthetic thinking requires several disciplined practices:

Methodological Rigor: Each perspective included in the analysis must meet standards of evidence and reasoning appropriate to its domain. Personal anecdotes, cultural myths, and unfounded speculation cannot be placed on equal footing with peer-reviewed research, controlled experiments, or systematic observation.

Explicit Boundary Recognition: Every framework has limitations—domains where it applies well and others where it fails. Physics provides excellent models for mechanical systems but poor guidance for social dynamics. Psychology illuminates individual behavior but struggles with quantum phenomena. Acknowledging these boundaries prevents inappropriate extrapolation.

Convergent Validation: The most compelling insights emerge when independent methods point toward similar conclusions. If neuroscience, anthropology, and phenomenology all suggest that consciousness has specific structural features, this convergence deserves serious attention. If only one approach supports a particular claim, skepticism is warranted.

Productive Contradiction Management: When different approaches yield incompatible conclusions, Venn Thinking treats these contradictions as information rather than problems to be eliminated. Contradictions often reveal hidden assumptions, methodological limitations, or the need for more sophisticated theoretical frameworks.

This approach draws on established principles from mixed-methods research in psychology and sociology, convergent validity testing in measurement theory, and triangulation strategies in anthropological fieldwork.[6] The goal is not to eliminate disagreement but to ensure that disagreements are productive—that they advance understanding rather than merely reflecting disciplinary prejudices.

Principles of Venn Thinking

Based on our exploration throughout this book, we can articulate five core principles for applying Venn Thinking to anomalous phenomena:

1. Start with Multiple Valid Maps Choose perspectives that have genuine scholarly grounding, even when they contradict each other. A valid map in this context means an approach that: (a) uses systematic methods appropriate to its domain, (b) acknowledges its own limitations, (c) makes falsifiable claims when possible, and (d) engages seriously with contrary evidence.

For investigating near-death experiences, valid maps might include cardiac resuscitation research, neuroscience of dying brains, cross-cultural thanatology, psychology of altered states, and phenomenological analysis of reported experiences. Invalid maps would include unfounded speculation, cherry-picked anecdotes, or approaches that dismiss contrary evidence without engagement.

2. Find the Overlap Zones Look for shared observations, repeatable patterns, or parallel insights across different approaches. These zones of convergence often reveal robust features of phenomena that transcend particular theoretical commitments.

When investigating telepathy, for instance, we might notice that laboratory parapsychology, crisis intervention psychology, and anthropological studies of healing practices all document apparent information transfer under specific conditions—emotional crisis, deep interpersonal bonds, and altered states of consciousness. This convergence doesn't prove telepathy exists, but it identifies conditions worth investigating more carefully.

3. Respect the Differences The goal is not to force artificial agreement but to let genuine differences sharpen our questions. Where perspectives diverge, we often discover the most interesting problems.

Neuroscience and Buddhist contemplative traditions offer different accounts of the self—one emphasizing neural networks, the other emphasizing the illusory nature of selfhood. Rather than trying to reconcile these views, Venn Thinking asks: What would selfhood look like if both perspectives captured important aspects of the phenomenon? How might neural network activity relate to experiences of self-dissolution?

4. Include the Cultural Lens Every phenomenon is filtered through human meaning-making systems. Cultural frameworks shape not only how experiences are interpreted but often how they are initially perceived and reported.

The same neurophysiological state might be experienced as "divine union" in a Christian context, "enlightenment" in a Buddhist framework, "cosmic consciousness" in a New Age setting, or "temporal lobe epilepsy" in a medical context. Venn Thinking treats these different framings as data about both the underlying phenomenon and the cultural systems that shape human experience.

5. Stay Provisional Overlap zones generate hypotheses, not final answers. The goal is to identify promising directions for investigation, not to settle debates definitively.

When multiple approaches suggest that consciousness might have non-local properties—quantum mechanics through entanglement, neuroscience through field effects, psychology through collective behavior—this convergence warrants serious investigation. But it doesn't prove that consciousness is quantum mechanical or that telepathy is real.

Case Study: Applying Venn Thinking to Near-Death Experiences

Let's examine how Venn Thinking applies to one of the phenomena we explored in detail:

Neuroscience Perspective: During cardiac arrest, dying brains exhibit burst patterns of electrical activity that could generate vivid, meaningful experiences. Decreased blood flow to specific brain regions might produce out-of-body sensations, while neurochemical changes could generate feelings of peace and encounters with deceased relatives.

Medical Research Perspective: Cardiologists and emergency physicians document cases where patients report accurate perceptions of resuscitation procedures while clinically unconscious. Some of these reports include details that are difficult to explain through normal sensory channels or prior knowledge.

Cross-Cultural Perspective: Anthropologists document remarkably consistent features of near-death experiences across diverse cultures—tunnels of light, life reviews, encounters with deceased relatives, decisions about returning to life. These consistencies suggest either universal neurological patterns or genuine encounters with non-ordinary realities.

Psychological Perspective: Researchers in trauma and altered states note that NDEs often occur during extreme stress, when normal psychological defenses are overwhelmed. The experiences might represent adaptive responses to life-threatening situations—providing comfort, meaning, and motivation for survival.

Phenomenological Perspective: Detailed analysis of first-person reports reveals complex, coherent experiences that don't easily fit standard models of hallucination or confabulation. The experiences often transform individuals' worldviews in lasting ways that suggest profound authenticity to the experients.

The Venn Zone: Where Perspectives Intersect

When we overlay these perspectives, several intriguing patterns emerge:

Convergence on Altered States: All approaches agree that NDEs occur during radically altered physiological and psychological states. This convergence suggests that understanding these alterations—whether through neuroscience, psychology, or contemplative practice—is crucial for explaining the phenomenon.

The Veridical Perception Puzzle: Medical and neuroscience perspectives struggle to explain reports of accurate perception during unconsciousness, while cultural and phenomenological approaches take these reports seriously. This tension highlights the need for better monitoring of brain activity during cardiac arrest and more rigorous testing of perceptual claims.

Transformation and Meaning: Psychological and cultural approaches emphasize the profound life changes that often follow NDEs, while neuroscience struggles to explain why brief brain states would have such lasting effects. This divergence suggests that meaning-making processes might be as important as neurophysiological mechanisms.

Universal Patterns and Cultural Variations: The consistent features across cultures (noted by anthropology) combined with specific cultural imagery (documented by phenomenology) suggest that NDEs might involve both universal neurological processes and culturally shaped interpretive frameworks.

These convergences and divergences don't resolve the debate about what NDEs represent, but they identify specific areas where focused research might be most productive. They also reveal assumptions that each approach brings to the investigation.

The Power of Productive Contradictions

In Venn Thinking, contradictions are not obstacles to be eliminated but pressure points that reveal the edges of our understanding. Where frameworks clash most dramatically, there is often the greatest potential for new insight.

Consider the tension between quantum mechanics and neuroscience approaches to consciousness. Quantum theories propose that consciousness involves non-local, entangled processes that transcend classical brain activity. Neuroscience insists that consciousness emerges from local, classical neural networks.

Rather than choosing sides, Venn Thinking asks: What would consciousness look like if both perspectives captured important aspects? This question has generated research into quantum biology, investigations of coherent quantum processes in neural microtubules, and exploration of how classical neural networks might interface with quantum information processing.[7]

Similarly, the contradiction between materialist and dualist approaches to psychedelic experiences—are DMT entities hallucinations or encounters?—pushes us to examine fundamental assumptions about the nature of reality and perception. If subjective experiences feel completely real to the experient, what does "reality" mean? How do we distinguish between generated and encountered phenomena?

These contradictions prevent premature closure while identifying areas where new theoretical frameworks might be needed. They also reveal the philosophical assumptions embedded in different research programs.

Methodological Challenges and Limitations

Venn Thinking faces several significant challenges that must be acknowledged:

Complexity and Ambiguity: Integrating multiple perspectives often increases rather than decreases complexity. The resulting picture may be more accurate but less immediately actionable than single-framework approaches. This can frustrate those seeking clear, definitive answers.

Boundary Problems: Determining which perspectives deserve inclusion in the analysis requires careful judgment. The approach must remain open to diverse viewpoints while maintaining standards of evidence and reasoning.

Communication Difficulties: Translating insights across disciplinary boundaries requires mastering multiple technical vocabularies and methodological assumptions. This can make synthetic work less accessible to specialists within particular fields.

Institutional Barriers: Academic reward systems favor depth within disciplines over breadth across them. Researchers who pursue synthetic approaches may find themselves without a clear intellectual home.

Philosopher Thomas Nagel has argued that some phenomena may be inherently resistant to unified explanation—that different aspects require fundamentally different types of understanding.[8] Mary Douglas has noted that institutional structures often prevent genuine interdisciplinary collaboration, leading to superficial exchanges rather than deep integration.[9]

These challenges suggest that Venn Thinking is most appropriately applied to phenomena that are genuinely complex and multi-faceted, where single-framework approaches have reached their limits. It also suggests the need for institutional support for interdisciplinary research and training programs that prepare researchers to work across boundaries.

Practical Applications Beyond Anomalous Phenomena

The principles of Venn Thinking extend far beyond the investigation of controversial topics. The approach proves valuable wherever complex phenomena require integration of multiple knowledge systems:

Climate Science: Understanding global climate change requires integrating atmospheric physics, oceanography, ecology, economics, political science, and human psychology. No single discipline can capture the full complexity of climate systems or develop effective policy responses.

Public Health: Addressing epidemics like COVID-19 requires coordination between virology, epidemiology, economics, psychology, political science, and anthropology. Purely medical approaches that ignore social and economic factors consistently fail to achieve public health goals.

Artificial Intelligence Development: Creating beneficial AI systems requires integration of computer science, cognitive psychology, philosophy of mind, ethics, sociology, and economics. Technical development without attention to social implications has repeatedly generated unintended consequences.

Personal Decision-Making: Individuals facing complex choices—career changes, relationship decisions, health treatments—benefit from considering multiple perspectives: emotional intuition, rational analysis, cultural values, spiritual principles, and practical constraints.

In each case, the goal is not to eliminate disagreement between different approaches but to extract insights from their intersection while respecting their individual contributions.

A Framework for Practice

For readers interested in applying Venn Thinking to their own investigations, whether academic or personal, here are key questions to guide the process:

Mapping Phase: What are the major perspectives that address this phenomenon? Which of these perspectives meet standards of rigor appropriate to their domains? What are the strengths and limitations of each approach?

Intersection Phase: Where do different perspectives agree on key observations or patterns? Where do they disagree, and what accounts for these disagreements? What assumptions does each perspective bring to the investigation?

Synthesis Phase: What new questions emerge from considering multiple perspectives simultaneously? What research directions might be most productive for advancing understanding? How might apparent contradictions point toward more sophisticated theoretical frameworks?

Application Phase: How might insights from this synthetic analysis inform practical decisions or future investigations? What remains genuinely uncertain, and how should this uncertainty influence action?

This framework prevents both premature closure and paralytic indecision. It enables action based on the best available understanding while maintaining appropriate humility about the limits of that understanding.

Venn Thinking and Future Research

The phenomena explored in this book—consciousness, anomalous experiences, the relationship between mind and reality—are likely to remain active areas of investigation for decades to come. As our theoretical frameworks and methodological tools continue to evolve, Venn Thinking provides a stable approach for integrating new insights with existing knowledge.

Consider how emerging technologies might transform our understanding of consciousness: brain-computer interfaces that allow direct monitoring of neural activity, virtual reality systems that can induce specific altered states, artificial intelligence systems that exhibit apparent creativity and self-awareness. Each of these developments will generate new data and theoretical possibilities that must be integrated with existing knowledge from neuroscience, psychology, philosophy, and contemplative traditions.

Similarly, as our understanding of quantum mechanics, network dynamics, and biological information processing continues to advance, new theoretical frameworks may emerge that can better accommodate anomalous phenomena. Venn Thinking provides a method for evaluating these frameworks against multiple types of evidence rather than single-perspective criteria.

The approach also suggests strategies for productive collaboration between researchers with different theoretical commitments. Rather than trying to convert others to particular worldviews, researchers can focus on identifying shared empirical questions and designing studies that provide meaningful tests across different theoretical frameworks.

The Limits of Knowledge and the Value of Mystery

Venn Thinking does not promise to resolve all mysteries or eliminate all uncertainties. Indeed, one of its most important functions is revealing the genuine limits of current knowledge—identifying areas where we simply don't yet have adequate theoretical frameworks or empirical tools.

This recognition of limits is not a failure but a crucial form of intellectual honesty. It prevents overconfidence in partial explanations while directing attention toward areas where new insights are most needed. It also maintains space for genuine wonder—the recognition that reality is more complex and interesting than any current theoretical framework can fully capture.

The history of science suggests that our most confident explanations often turn out to be provisional. Classical physics seemed complete until quantum mechanics revealed its limitations. Newtonian gravity appeared universal until general relativity provided a more comprehensive framework. Genetic determinism dominated biology until epigenetics revealed the complexity of gene expression.

This pattern suggests that our current understanding of consciousness, reality, and anomalous phenomena is also likely to be revised as new evidence emerges and new theoretical frameworks develop. Venn Thinking provides a method for remaining open to these revisions while extracting maximum insight from current knowledge.

Beyond the Interface

If Part I provided us with theoretical tools and Part II applied these tools to specific phenomena, Part III offers a method for continued investigation as our knowledge continues to evolve. Venn Thinking is not a final answer but a framework for asking better questions.

The title of this book—*Beyond the Interface*—suggests that many of the phenomena we've explored occur at the boundaries between different domains: between mind and world, between individual and collective consciousness, between ordinary and extraordinary states of awareness. These boundary regions are precisely where Venn Thinking proves most valuable, where the intersection of multiple perspectives reveals patterns invisible to any single approach.

As we continue to probe these boundaries, we need tools that are simultaneously rigorous and flexible, that can accommodate both skeptical analysis and genuine openness to surprise. We need methods that respect the complexity of phenomena while remaining grounded in empirical observation and logical reasoning.

The cartographers of the 16th century eventually produced maps that enabled reliable navigation across oceans. But they never completed the mapping process—new coastlines, channels, and islands continue to be discovered. Similarly, our investigation of consciousness and reality is an ongoing voyage rather than a destination.

The value lies not in arriving at final answers but in developing better maps—more accurate, more comprehensive, more honest about what remains unknown. Venn Thinking provides a compass for this continuing journey, helping us navigate the intersection of the known and the mysterious with both rigor and wonder.

When multiple maps converge on the same coastline, we can be reasonably confident it exists. When they diverge, we know exactly where the mystery still lies—and where to send the next expedition.

Practicing Venn Thinking: A Guide for Inquiry

When confronting any complex phenomenon or decision:

1. **Identify multiple valid perspectives** that address the question using rigorous methods appropriate to their domains
2. **Map the intersections** where different approaches agree on key observations or reach similar conclusions
3. **Examine the divergences** where approaches disagree, asking what accounts for these differences
4. **Look for productive contradictions** that reveal assumptions or point toward new questions

5. **Include cultural and contextual factors** that shape how the phenomenon is experienced and interpreted
6. **Generate new hypotheses** based on insights that emerge from intersection analysis
7. **Maintain provisional stance** toward conclusions while remaining open to revision based on new evidence
8. **Identify areas of genuine uncertainty** where current knowledge remains inadequate
9. **Direct future inquiry** toward areas where synthetic analysis suggests the highest potential for new insight

Remember: The goal is not to force agreement between different perspectives but to extract maximum insight from their intersection while respecting their individual contributions.

Afterword: Wonder as Method

In December 1972, the crew of Apollo 17 took what became known as the "Blue Marble" photograph—the first full-color image of Earth from space captured by human hands. Suspended in the cosmic blackness, our planet appeared at once utterly fragile and magnificently whole. No political borders were visible, no human divisions apparent, just one shimmering sphere of blue ocean and brown land beneath swirling white clouds.

That single image did not answer any scientific question. It settled no theoretical debate, proved no hypothesis, resolved no technical dispute. Yet it fundamentally transformed how millions of people understood their place in the universe. It shifted perspectives not through data or argument, but through the immediate, visceral experience of awe.

The photograph sparked the modern environmental movement, influenced space policy for decades, and continues to shape how we think about planetary stewardship. It demonstrated something crucial about the relationship between wonder and knowledge: sometimes the most profound insights come not from accumulating facts but from seeing familiar things with fresh eyes.

Awe as the Engine of Discovery

Throughout the history of science, wonder has served as the ignition point for transformative discoveries. When Benjamin Franklin watched lightning illuminate the Philadelphia sky, his sense of awe at the phenomenon drove him to conduct dangerous experiments that revealed the electrical nature of atmospheric discharge. When Barbara McClintock observed the intricate patterns of genetic inheritance in corn plants, her wonder at the complexity of biological systems led her to discover genetic transposition—"jumping genes"—decades before the scientific community was ready to accept her findings.[1]

Charles Darwin's theory of evolution emerged from his sense of amazement at the diversity of life forms he encountered during the voyage of the Beagle. As he wrote in his journal, "It is impossible to reflect on the changed state of the American continent without the deepest astonishment."[2] That astonishment drove decades of careful observation, experimentation, and theoretical development.

More recently, Jocelyn Bell Burnell's discovery of pulsars began with her wonder at strange radio signals that didn't fit existing models of cosmic phenomena. Rather than dismissing the signals as equipment malfunction—as many of her colleagues initially suggested—she followed her curiosity and uncovered an entirely new class of stellar objects.[3]

Even Albert Einstein, perhaps the most celebrated theoretical physicist in history, emphasized the central role of wonder in

304

scientific work. "The most beautiful thing we can experience is the mysterious," he wrote. "It is the source of all true art and science."[4]

These examples share a crucial pattern: the scientists who made revolutionary discoveries maintained what psychologist Dacher Keltner calls "epistemic awe"—wonder specifically directed toward expanding understanding.[5] They approached mysteries not with the goal of quickly explaining them away, but with genuine curiosity about what those mysteries might reveal.

Wonder in the Phenomena We've Explored

This dynamic relationship between wonder and rigorous investigation runs throughout the phenomena we've examined in this book. Consider how wonder has driven research in each domain:

Near-death experiences captured scientific attention precisely because they seemed so extraordinary. Researchers like Raymond Moody, Bruce Greyson, and Pim van Lommel were initially struck by the profound nature of reported experiences during cardiac arrest. Their sense of wonder at these accounts motivated decades of careful data collection, methodological refinement, and theoretical development. Whether NDEs ultimately prove to involve consciousness beyond the brain or represent remarkable neurological phenomena, the research has advanced our understanding of consciousness during extreme states.

Reincarnation research began with Ian Stevenson's amazement at young children's detailed reports of previous lives. His wonder at these accounts drove him to develop rigorous investigative methods, travel extensively to document cases, and maintain careful records over decades. While debates continue about interpretation, his work established standards for investigating claims about personal identity and memory that extend far beyond reincarnation research.

Precognition studies emerged from researchers' fascination with reports of apparent future knowledge. Dean Radin's wonder at statistical anomalies in consciousness research motivated sophisticated meta-analyses and new experimental paradigms. Whether these studies reveal genuine precognitive abilities or subtle methodological artifacts, they have advanced our understanding of time perception, probability assessment, and the psychology of intuitive decision-making.

Telepathy research has been sustained by wonder at apparent mind-to-mind communication. While remaining highly controversial, investigations of telepathic phenomena have contributed to our understanding of unconscious social coordination, emotional contagion, and the psychology of interpersonal connection.

Psychedelic research has resurged largely due to researchers' awe at the profound, often life-changing experiences induced by these substances. Scientists like Robin Carhart-Harris, Roland Griffiths, and Matthew Johnson have combined genuine wonder at the phenomena with rigorous methodology, leading to breakthrough insights about consciousness, neural plasticity, and mental health treatment.

Savant abilities continue to fascinate researchers because they reveal unexpected dimensions of human cognitive potential. Wonder at extraordinary abilities in individuals with developmental disabilities has driven advances in our understanding of brain plasticity, memory systems, and the relationship between consciousness and cognitive architecture.

In each case, wonder provided the motivational energy for sustained investigation while rigorous methodology ensured that investigation remained productive. The combination proved essential—wonder without rigor leads to speculation; rigor without wonder often misses the most interesting questions.

The Science of Wonder Itself

Recent research has begun to illuminate wonder as a phenomenon worthy of scientific investigation in its own right. Neuroscience studies reveal that experiences of awe activate specific brain networks associated with attention, self-transcendence, and cognitive flexibility.[6] Psychology research demonstrates that awe experiences increase openness to new ideas, enhance creative problem-solving, and reduce dogmatic thinking.[7]

These findings suggest that wonder is not merely a pleasant emotional experience but a cognitive state that enhances our capacity for discovery. When we experience genuine awe, our brains become more receptive to information that challenges existing assumptions and more capable of generating novel connections between disparate ideas.

This research validates what many scientists have long suspected: wonder is not opposed to rigorous thinking but essential to it. The capacity for awe keeps us mentally flexible enough to recognize when our current theories are inadequate and our existing methods insufficient.

Wonder Without Naïveté

One of the most important lessons of this book is that wonder need not require the suspension of critical thinking. Genuine awe can coexist with healthy skepticism, careful methodology, and rigorous analysis. In fact, the most productive research often emerges from investigators who combine deep curiosity with exacting standards of evidence.

Consider how researchers have approached the phenomena in Part II. The best work in each field demonstrates sophisticated appreciation for methodological challenges, cultural influences, and alternative explanations. Pim van Lommel's near-death experience research includes careful control groups and consideration of neurological alternatives. Jim Tucker's reincarnation investigations incorporate stringent verification procedures and analysis of cultural factors. Dean Radin's consciousness research employs sophisticated statistical techniques and addresses multiple explanatory hypotheses.

This approach—wonder coupled with methodological sophistication—represents what we might call "mature awe." It maintains openness to extraordinary possibilities while insisting on extraordinary evidence. It treats mysteries as invitations to investigation rather than problems to be quickly dismissed or uncritically accepted.

Such mature awe requires what philosopher Paul Ricoeur called a "second naïveté"—the ability to experience wonder again after passing through critical analysis.[8] This is not the simple naïveté of those who have never questioned their assumptions, but the earned naïveté of those who have examined their beliefs carefully and chosen to remain open to mystery.

Current Debates: Wonder in Contemporary Science

The relationship between wonder and rigorous investigation has become a topic of explicit discussion in several scientific fields. Some researchers argue that contemporary science has become too conservative, too focused on incremental advances within established paradigms, and insufficiently open to transformative discoveries.[9]

In physics, researchers debate whether string theory's mathematical elegance justifies continued investigation despite lack of empirical confirmation. The sense of wonder at the theory's potential for unifying fundamental forces motivates some researchers while others argue for focusing on more empirically tractable questions.

In consciousness studies, researchers grapple with how to maintain scientific rigor while investigating phenomena that may require new theoretical frameworks. The field struggles with balancing openness to transformative discoveries with appropriate skepticism about extraordinary claims.

In psychology and neuroscience, researchers debate how to study subjective experiences—from mystical states to psychedelic experiences to reports of anomalous phenomena—that resist easy quantification but may provide important insights into the nature of consciousness.

These debates reflect a broader tension in contemporary science between the need for rigorous methodology and the recognition that some of the most important discoveries may require venturing beyond established paradigms. The resolution, we suggest, lies not in choosing between wonder and rigor but in developing approaches that integrate both.

Venn Thinking as Structured Wonder

The Venn Thinking approach developed in Chapter 14 provides a framework for maintaining wonder while ensuring rigorous investigation. By systematically examining phenomena from multiple theoretical perspectives, we can remain open to extraordinary possibilities while subjecting those possibilities to comprehensive analysis.

This approach recognizes that wonder often emerges at the intersections between different ways of understanding reality. When neuroscience, anthropology, and phenomenology all point toward interesting patterns in mystical experience, our sense of awe is enhanced rather than diminished. When physics, psychology, and consciousness research converge on questions about the nature of time and causality, our wonder at the complexity of reality deepens.

Venn Thinking thus provides a method for scaling wonder—for experiencing awe not just at individual phenomena but at the intricate patterns that emerge when multiple rigorous approaches are brought into conversation. It transforms wonder from a purely emotional response into a systematic tool for discovery.

Living in the Questions

Perhaps the most important insight from our investigation is that the goal of inquiry is not to eliminate mystery but to engage with it more skillfully. The phenomena explored in Part II— from near-death experiences to savant abilities—remain genuinely puzzling despite decades of research. But our puzzlement is now more informed, more nuanced, and more productive.

We know more about what questions to ask, what methods might prove fruitful, and what theoretical frameworks offer the most promise. We understand better how cultural factors shape experience and interpretation, how methodological choices influence results, and how different explanatory approaches illuminate different aspects of complex phenomena.

This enhanced understanding does not resolve the mysteries, but it transforms our relationship to them. We can live more comfortably with uncertainty while remaining actively engaged with investigation. We can maintain wonder while exercising critical judgment. We can hold multiple possibilities simultaneously without falling into either dogmatic belief or premature dismissal.

The poet Rainer Maria Rilke captured this state beautifully in his *Letters to a Young Poet*: "Try to love the questions themselves... Do not now seek the answers, which cannot be given you because you would not be able to live them. And the point is, to live everything. Live the questions now. Perhaps you will then gradually, without noticing it, live along some distant day into the answer."[10]

Personal Applications: Wonder as Intellectual Practice

For readers seeking to apply these insights beyond the specific phenomena discussed in this book, wonder can become a deliberate intellectual practice. This involves cultivating several related capacities:

Maintaining beginner's mind: Approaching familiar topics with fresh curiosity, questioning assumptions that may have become invisible through habitual thinking.

Embracing productive uncertainty: Learning to sit comfortably with questions that lack immediate answers while remaining actively engaged with investigation.

Practicing perspective-taking: Deliberately examining issues from multiple theoretical, cultural, and methodological standpoints to reveal hidden assumptions and generate new insights.

Balancing openness and skepticism: Remaining receptive to surprising possibilities while maintaining appropriate standards of evidence and reasoning.

Cultivating awe for complexity: Developing appreciation for the intricate, multifaceted nature of reality rather than seeking oversimplified explanations.

These practices prove valuable far beyond the investigation of anomalous phenomena. They enhance creativity in artistic pursuits, deepen understanding in academic disciplines, improve decision-making in complex professional situations, and enrich personal relationships by maintaining curiosity about other people's experiences and perspectives.

Looking Forward: The Future of Wonder-Driven Inquiry

As we advance further into the 21st century, the integration of wonder and rigorous methodology may become increasingly important. We face challenges—from climate change to artificial intelligence to global social coordination—that require both innovative thinking and careful analysis. Solutions are likely to emerge from the intersection of multiple disciplines, demanding the kind of synthetic thinking that wonder helps facilitate.

The phenomena explored in this book may serve as a training ground for developing these capacities. Learning to think rigorously about consciousness, anomalous experiences, and the relationship between mind and reality exercises exactly the intellectual muscles needed for addressing other complex challenges.

Moreover, some of the theoretical frameworks we've explored—from quantum mechanics to network science to predictive processing—may prove relevant to understanding and addressing broader issues. The principles of emergence and self-organization that appear in consciousness research also illuminate social dynamics and technological development. The insights about uncertainty and complexity that arise from studying anomalous phenomena apply equally to policy-making and strategic planning.

The Ongoing Expedition

The cave metaphor that opened Chapter 13 provides a fitting framework for understanding our ongoing relationship with mystery. Each generation of explorers maps new chambers, develops better tools, and discovers previously unknown passages. But the cave continues to extend beyond current exploration, promising new discoveries for future investigators.

The same pattern characterizes our investigation of consciousness and reality. Each generation of researchers advances understanding, develops more sophisticated methods, and asks more precise questions. But the fundamental mysteries remain, continuing to inspire wonder and drive inquiry.

This perspective transforms our relationship to the unknown. Rather than viewing unresolved questions as failures of scientific progress, we can appreciate them as invitations to continued discovery. Rather than demanding immediate answers to profound mysteries, we can enjoy the process of investigation itself.

Whether you are a professional scientist, a student, a policy-maker, or simply a curious human being, you have a role in this ongoing expedition. Every time you approach a mystery with both wonder and rigor, every time you consider multiple perspectives on a complex issue, every time you remain open to surprising possibilities while exercising critical judgment, you contribute to humanity's collective effort to understand reality more deeply.

The Deeper Method

Throughout this book, we have provided specific tools and frameworks: predictive processing models, consciousness theories, quantum mechanics principles, network dynamics, cultural analysis methods, and Venn Thinking approaches. These tools are valuable and we encourage their continued use and development.

But beneath all these specific methods lies something more fundamental: the cultivation of wonder as an intellectual virtue. This involves not just feeling awe at mysterious phenomena, but developing the capacity to let that awe guide us toward more rigorous investigation, more sophisticated questions, and more nuanced understanding.

Wonder, in this sense, is not merely an emotional response but a cognitive stance—a way of approaching reality that remains open to surprise while insisting on careful analysis. It is a method for staying mentally flexible in an complex world, for maintaining intellectual humility in the face of vast unknowns, and for finding joy in the process of discovery itself.

This deeper method cannot be reduced to a set of procedures or protocols. It must be cultivated through practice, refined through experience, and renewed through continued encounter with mystery. It requires both intellectual sophistication and emotional openness, both critical thinking and creative imagination.

Final Reflection

One day, many of the phenomena explored in Part II may have clear, widely accepted explanations. Perhaps we will discover that near-death experiences result entirely from specific neurochemical processes during brain hypoxia. Perhaps reincarnation reports will be fully explained through cultural transmission and memory reconstruction. Perhaps telepathy and precognition will prove to be statistical artifacts and confirmation bias.

Or perhaps we will find that consciousness truly does extend beyond individual brains, that information can travel backwards through time, that reality is far stranger and more wonderful than current scientific paradigms suggest.

Whatever the eventual outcomes, what will matter most is not whether we were "right" in our current speculations, but whether we maintained the combination of wonder and rigor that makes genuine discovery possible. What will matter is whether we kept exploring, kept questioning, kept holding space for complexity and mystery while pursuing understanding with appropriate care and sophistication.

The phenomena discussed in this book may ultimately prove to be windows into extraordinary aspects of reality, or they may reveal the remarkable creativity of human consciousness in constructing meaningful experiences from ambiguous inputs. Either discovery would be profound. Either would deepen our appreciation for the intricate relationship between mind and world.

In the end, the real method is not any particular technique or theoretical framework, though these remain important tools. The real method is the cultivation of wonder itself—the capacity to approach reality with genuine curiosity, appropriate humility, and sustained commitment to understanding whatever we might discover.

Because the greatest discovery of all may be learning to live skillfully with mystery, finding joy in questions as well as answers, and maintaining awe for the profound complexity of existence itself.

The Blue Marble photograph showed us Earth as it truly is: a small, precious world suspended in vast emptiness, beautiful beyond our capacity to fully comprehend. In the same way, our investigation of consciousness and reality reveals something equally remarkable: the universe is stranger, more intricate, and more wonderful than any of our current theories can fully capture.

That recognition is not a limit to knowledge but its greatest achievement—the discovery that reality exceeds our current capacity to understand it completely, and therefore promises endless opportunities for continued wonder, investigation, and discovery.

The expedition continues. The next chamber awaits. And wonder lights the way.

Notes by Chapter

Introduction: Through the Interface

1. Pim van Lommel et al., "Near-death experience in survivors of cardiac arrest: a prospective study in the Netherlands," *The Lancet* 358, no. 9298 (2001): 2039-2045. Van Lommel clarifies that "clinical death" refers to the period when heartbeat and blood pressure are undetectable, distinguishing this from brain death.

2. Andrew N. Jordan et al., "Quantum Delayed-Choice Experiments," *Reviews of Modern Physics* 88, no. 1 (2016): 015005. See also Alain Aspect, "Bell's inequality test: more ideal than ever," *Nature* 398 (1999): 189-190.

3. Andrea Cavagna et al., "Scale-Free Correlations in Starling Flocks," *Proceedings of the National Academy of Sciences* 107, no. 26 (2010): 11865-11870.

4. Karl Friston, "The free-energy principle: a unified brain theory?" *Nature Reviews Neuroscience* 11, no. 2 (2010): 127-138.

5. Andy Clark, "Whatever next? Predictive brains, situated agents, and the future of cognitive science," *Behavioral and Brain Sciences* 36, no. 3 (2013): 181-204.

6. Lisa Feldman Barrett, *How Emotions Are Made: The Secret Life of the Brain* (Boston: Houghton Mifflin Harcourt, 2017).

Prologue: What Was Once Unexplainable

1. Benjamin Franklin, "Experiments and Observations on Electricity" (Philadelphia: E. Cave, 1751); I. Bernard Cohen, *Benjamin Franklin's Science* (Cambridge, MA: Harvard University Press, 1990).

2. Louis Pasteur, "The Germ Theory and Its Applications to Medicine and Surgery," *Comptes Rendus de l'Académie des Sciences* 86 (1878): 1037-1043; Robert Koch, "The Etiology of Tuberculosis," *Berliner Klinische Wochenschrift* 19 (1882): 221-230.

3. Isaac Newton, *Philosophiæ Naturalis Principia Mathematica* (London: Joseph Streater, 1687); Albert Einstein, "The Foundation of the General Theory of Relativity," *Annalen der Physik* 49, no. 7 (1916): 769-822.

4. William Thomson (Lord Kelvin), "On the Age of the Sun's Heat," *Macmillan's Magazine* 5 (1862): 388-393; Tom D. Crouch, *The Bishop's Boys: A Life of Wilbur and Orville Wright* (New York: W. W. Norton, 1989).

5. Joseph Needham, *Science and Civilisation in China*, vol. 4, "Physics and Physical Technology" (Cambridge: Cambridge University Press, 1962).

Chapter 1: The Predictive Brain

1. Bannert, Maria M., and Andreas Bartels. "Decoding the Yellow of a Gray Banana." *Current Biology* 23, no. 22 (2013): 2268-2272.

2. Samuel, Arthur G. "Phonemic restoration: insights from a new methodology." *Journal of Experimental Psychology: General* 110, no. 4 (1981): 474-494.

3. Land, Michael F., and David N. Lee. "Where we look when we steer." *Nature* 369, no. 6483 (1994): 742-744.

4. Friston, Karl. "The free-energy principle: a unified brain theory?" *Nature Reviews Neuroscience* 11, no. 2 (2010): 127-138.

5. Clark, Andy. *Surfing Uncertainty: Prediction, Action, and the Embodied Mind* (Oxford: Oxford University Press, 2015).

6. Libet, Benjamin, et al. "Time of conscious intention to act in relation to onset of cerebral activity (readiness-potential): the unconscious initiation of a freely voluntary act." *Brain* 106, no. 3 (1983): 623-642.

7. Benedetti, Fabrizio, et al. "Neurobiological mechanisms of the placebo effect." *Journal of Neuroscience* 25, no. 45 (2005): 10390-10402.

8. Oken, Barry S. "Placebo effects: clinical aspects and neurobiology." *Brain* 131, no. 11 (2008): 2812-2823.

9. Häuser, Winfried, et al. "Nocebo phenomena in medicine: their relevance in everyday clinical practice." *Deutsches Ärzteblatt International* 109, no. 26 (2012): 459-465.

10. Friston, Karl. "A theory of cortical responses." *Philosophical Transactions of the Royal Society B* 360, no. 1456 (2005): 815-836.

11. Seth, Anil K. "Interoceptive inference, emotion, and the embodied self." *Trends in Cognitive Sciences* 17, no. 11 (2013): 565-573.

12. Clark, Andy. "Whatever next? Predictive brains, situated agents, and the future of cognitive science." *Behavioral and Brain Sciences* 36, no. 3 (2013): 181-204.

13. Hohwy, Jakob. *The Predictive Mind: Cognitive Science and the Future of Knowledge* (Oxford: Oxford University Press, 2013).

14. Barrett, Lisa Feldman. "The theory of constructed emotion: an active inference account of interoception and categorization." *Social Cognitive and Affective Neuroscience* 12, no. 1 (2017): 1-23.

15. Bowers, Jeffrey S., and Colin J. Davis. "Bayesian just-so stories in psychology and neuroscience." *Psychological Bulletin* 138, no. 3 (2012): 389-414.

16. de Lange, Floris P., et al. "How do expectations shape perception?" *Trends in Cognitive Sciences* 22, no. 9 (2018): 764-779.

Chapter 2: Consciousness—Fundamental or Emergent?

1. Olaf Blanke et al., "Stimulating illusory own-body perceptions," *Nature* 419, no. 6904 (2002): 269-270.

2. David J. Chalmers, "Facing up to the problem of consciousness," *Journal of Consciousness Studies* 2, no. 3 (1995): 200-219.

3. Giulio Tononi, "An information integration theory of consciousness," *BMC Neuroscience* 5, no. 1 (2004): 42.

4. Christof Koch, *The Feeling of Life Itself: Why Consciousness Is Widespread but Can't Be Computed* (Cambridge, MA: MIT Press, 2019).

5. Stanislas Dehaene, *Consciousness and the Brain: Deciphering How the Brain Codes Our Thoughts* (New York: Viking, 2014).

6. Michael S. Gazzaniga, "Forty-five years of split-brain research and still going strong," *Nature Reviews Neuroscience* 6, no. 8 (2005): 653-659.

7. Galen Strawson, "Consciousness and its place in nature," *Philosophical Studies* 120, no. 1-3 (2004): 185-204.

8. Philip Goff, *Galileo's Error: Foundations for a New Science of Consciousness* (New York: Pantheon, 2019).

9. David J. Chalmers, "Panpsychism and panprotopsychism," *Amherst Lecture in Philosophy* 8 (2013): 1-35.

10. Thomas Nagel, *Mind and Cosmos: Why the Materialist Neo-Darwinian Conception of Nature Is Almost Certainly False* (New York: Oxford University Press, 2012).

11. Antonio Damasio, *The Strange Order of Things: Life, Feeling, and the Making of Cultures* (New York: Pantheon, 2018).

12. Thomas Metzinger, *Being No One: The Self-Model Theory of Subjectivity* (Cambridge, MA: MIT Press, 2003).

13. Andy Clark and David Chalmers, "The extended mind," *Analysis* 58, no. 1 (1998): 7-19.

14. Anil Seth, "Consciousness as interoceptive inference," in *The Interoceptive Mind*, eds. Manos Tsakiris and Helena De Preester (Oxford: Oxford University Press, 2018), 51-71.

15. Stanislas Dehaene and Jean-Pierre Changeux, "Experimental and theoretical approaches to conscious processing," *Neuron* 70, no. 2 (2011): 200-227.

16. Daniel C. Dennett, *Consciousness Explained* (Boston: Little, Brown, 1991).

Chapter 3: Quantum Reality and Emergent Complexity

1. Andrew N. Jordan et al., "Quantum delayed-choice experiments," *Reviews of Modern Physics* 88, no. 1 (2016): 015005. The original experimental implementation was reported by Xin-Song Ma et al., "Quantum delayed-choice experiment," *Nature Physics* 8, no. 6 (2012): 479-484.

2. Alain Aspect, Philippe Grangier, and Gérard Roger, "Experimental realization of Einstein-Podolsky-Rosen-Bohm Gedankenexperiment: a new violation of Bell's inequalities," *Physical Review Letters* 49, no. 2 (1982): 91-94.

3. Sean Carroll, *Something Deeply Hidden: Quantum Worlds and the Emergence of Spacetime* (New York: Dutton, 2019).

4. Carlo Rovelli, "Relational quantum mechanics," *International Journal of Theoretical Physics* 35, no. 8 (1996): 1637-1678.

5. Christopher A. Fuchs, "QBism, the perimeter of quantum Bayesianism," *arXiv preprint arXiv:1003.5209* (2010).

6. Steven H. Strogatz, *Sync: How Order Emerges from Chaos in the Universe, Nature, and Daily Life* (New York: Hyperion, 2003).

7. Stuart A. Kauffman, *At Home in the Universe: The Search for Laws of Self-Organization and Complexity* (New York: Oxford University Press, 1995).

8. Edward N. Lorenz, "Deterministic nonperiodic flow," *Journal of Atmospheric Sciences* 20, no. 2 (1963): 130-141.

9. Michael Levin, "Bioelectric signaling: reprogrammable circuits underlying embryogenesis, regeneration, and cancer," *Cell* 184, no. 8 (2021): 1971-1989.

10. Eugene P. Wigner, "Remarks on the mind-body question," in *The Scientist Speculates*, ed. I.J. Good (London: Heinemann, 1961), 284-302.

11. Roger Penrose and Stuart Hameroff, "Consciousness in the universe: a review of the 'Orch OR' theory," *Physics of Life Reviews* 11, no. 1 (2014): 39-78.

12. Max Tegmark, "Importance of quantum decoherence in brain processes," *Physical Review E* 61, no. 4 (2000): 4194-4206.

13. Sabine Hossenfelder, *Existential Physics: A Scientist's Guide to Life's Biggest Questions* (New York: Viking, 2022).

14. Henry P. Stapp, *Mindful Universe: Quantum Mechanics and the Participating Observer* (Berlin: Springer, 2007).

15. Amit Goswami, *The Self-Aware Universe* (New York: Putnam, 1993).

16. Gregory D. Scholes et al., "Using coherence to enhance function in chemical and biophysical systems," *Nature* 543, no. 7647 (2017): 647-656.

Chapter 4: Artificial Minds as Cognitive Mirrors

1. This represents a typical example of ChatGPT's humor generation capabilities as documented by users since the model's release in November 2022. See OpenAI, "ChatGPT: Optimizing Language Models for Dialogue," https://openai.com/blog/chatgpt/.

2. Michael C. Frank et al., "Word learning as Bayesian inference," *Psychological Review* 116, no. 4 (2009): 953-973.

3. Yann LeCun, Yoshua Bengio, and Geoffrey Hinton, "Deep learning," *Nature* 521, no. 7553 (2015): 436-444.

4. Ashish Vaswani et al., "Attention is all you need," *Advances in Neural Information Processing Systems* 30 (2017): 5998-6008.

5. Karl Friston et al., "Active inference: a process theory," *Neural Computation* 29, no. 1 (2017): 1-49.

6. Michael Rubenstein et al., "Programmable self-assembly in a thousand-robot swarm," *Science* 345, no. 6198 (2014): 795-799.

7. Giulio Tononi, "Integrated information theory: from consciousness to its physical substrate," *Nature Reviews Neuroscience* 17, no. 7 (2016): 450-461.

8. Yoshua Bengio, "The consciousness prior," *arXiv preprint* arXiv:1709.08568 (2017).

9. Yann LeCun, "A path towards autonomous machine intelligence," *arXiv preprint* arXiv:2206.11031 (2022).

10. Joscha Bach, *Principles of Synthetic Intelligence: Psi: An Architecture of Motivated Cognition* (New York: Oxford University Press, 2009).

11. Melanie Mitchell, *Artificial Intelligence: A Guide for Thinking Humans* (New York: Farrar, Straus and Giroux, 2019).

12. John R. Searle, "Minds, brains, and programs," *Behavioral and Brain Sciences* 3, no. 3 (1980): 417-424.

13. Stuart Russell, *Human Compatible: Artificial Intelligence and the Problem of Control* (New York: Viking, 2019).

Chapter 5: Religion, Mysticism, and the Science of Subjective Realities

1. Richard Maurice Bucke, *Cosmic Consciousness: A Study in the Evolution of the Human Mind* (New York: E.P. Dutton, 1901).
2. Walter N. Pahnke, "Drugs and mysticism," *International Journal of Parapsychology* 8, no. 2 (1966): 295-313.
3. Robin L. Carhart-Harris et al., "Neural correlates of the psychedelic state as determined by fMRI studies with psilocybin," *Proceedings of the National Academy of Sciences* 109, no. 6 (2012): 2138-2143.
4. Roland R. Griffiths et al., "Psilocybin produces substantial and sustained decreases in depression and anxiety in patients with life-threatening cancer," *Journal of Psychopharmacology* 30, no. 12 (2016): 1181-1197.
5. Andrew B. Newberg et al., "The measurement of regional cerebral blood flow during the complex cognitive task of meditation," *Psychiatry Research: Neuroimaging* 106, no. 2 (2001): 113-122.
6. William James, *The Varieties of Religious Experience: A Study in Human Nature* (New York: Longmans, Green, and Co., 1902).
7. Stanislav Grof, *LSD: Doorway to the Numinous* (Rochester, VT: Park Street Press, 2009).
8. Erika Bourguignon, "A cross-cultural study of dissociational states," *Ohio State University Research Foundation* (1968).
9. Roland R. Griffiths et al., "Psilocybin can occasion mystical-type experiences having substantial and

sustained personal meaning and spiritual significance,"
Psychopharmacology 187, no. 3 (2006): 268-283.

10. Rick Strassman, *DMT: The Spirit Molecule* (Rochester, VT: Park Street Press, 2001).

11. Robin L. Carhart-Harris et al., "LSD enhances suggestibility in healthy volunteers," *Psychopharmacology* 232, no. 4 (2015): 785-794.

12. William James, *The Varieties of Religious Experience: A Study in Human Nature* (New York: Longmans, Green, and Co., 1902).

13. Andrew Newberg and Eugene D'Aquili, *Why God Won't Go Away: Brain Science and the Biology of Belief* (New York: Ballantine Books, 2001).

14. Robin L. Carhart-Harris, "How do psychedelics work?" *Current Opinion in Psychiatry* 32, no. 1 (2019): 16-21.

15. Roland R. Griffiths and William A. Richards, "Mystical experiences occasioned by the hallucinogen psilocybin lead to increases in the personality domain of openness," *Journal of Psychopharmacology* 25, no. 11 (2011): 1453-1461.

16. Stanislav Grof, *Psychology of the Future: Lessons from Modern Consciousness Research* (Albany: State University of New York Press, 2000).

17. Steven T. Katz, "Language, epistemology, and mysticism," in *Mysticism and Philosophical Analysis*, ed. Steven T. Katz (New York: Oxford University Press, 1978), 22-74.

Chapter 6: Systems and Networks—Patterns Behind the Unexplainable

1. Yuki Sugiyama et al., "Traffic jams without bottlenecks—experimental evidence for the physical mechanism of the formation of a jam," *New Journal of Physics* 10, no. 3 (2008): 033001.
2. Albert-László Barabási and Réka Albert, "Emergence of scaling in random networks," *Science* 286, no. 5439 (1999): 509-512.
3. Duncan J. Watts and Steven H. Strogatz, "Collective dynamics of 'small-world' networks," *Nature* 393, no. 6684 (1998): 440-442.
4. Steven H. Strogatz, *Sync: How Order Emerges from Chaos in the Universe, Nature, and Daily Life* (New York: Hyperion, 2003).
5. Olaf Sporns, *Networks of the Brain* (Cambridge, MA: MIT Press, 2010).
6. Deborah M. Gordon, *Ants at Work: How an Insect Society Is Organized* (New York: Free Press, 1999).
7. Michael Rubenstein et al., "Programmable self-assembly in a thousand-robot swarm," *Science* 345, no. 6198 (2014): 795-799.
8. Eva Jablonka and Marion J. Lamb, *Evolution in Four Dimensions: Genetic, Epigenetic, Behavioral, and Symbolic Variation in the History of Life* (Cambridge, MA: MIT Press, 2005).
9. Rachel Yehuda et al., "Holocaust exposure induced intergenerational effects on FKBP5 methylation," *Biological Psychiatry* 80, no. 5 (2016): 372-380.

10. Mark S. Granovetter, "The strength of weak ties," *American Journal of Sociology* 78, no. 6 (1973): 1360-1380.
11. Stanley Milgram, "The small world problem," *Psychology Today* 2, no. 1 (1967): 60-67.
12. Steven H. Strogatz, "Exploring complex networks," *Nature* 410, no. 6825 (2001): 268-276.
13. Albert-László Barabási, *Network Science* (Cambridge: Cambridge University Press, 2016).
14. Dirk Helbing, "Traffic and related self-driven many-particle systems," *Reviews of Modern Physics* 73, no. 4 (2001): 1067-1141.
15. Ed Bullmore and Olaf Sporns, "Complex brain networks: graph theoretical analysis of structural and functional systems," *Nature Reviews Neuroscience* 10, no. 3 (2009): 186-198.

Interlude: "Ok, You Obviously Wonder About Ghosts..."

1. For cross-cultural perspectives on ghost beliefs, see David Hufford, *The Terror That Comes in the Night: An Experience-Centered Study of Supernatural Assault Traditions* (Philadelphia: University of Pennsylvania Press, 1982); and Shirley Jackson Case, *Experience with the Supernatural in Early Christian Times* (New York: Century Co., 1929).
2. J. Allan Cheyne, "Sleep paralysis and the structure of waking-nightmare hallucinations," *Dreaming* 13, no. 3 (2003): 163-179.

3. Agneta Grimby, "Bereavement among elderly people: grief reactions, post-bereavement hallucinations and quality of life," *Acta Psychiatrica Scandinavica* 87, no. 1 (1993): 72-80.

4. For environmental factors in anomalous experiences, see Christopher C. French, "Fantastic memories: The relevance of research into eyewitness testimony and false memories for reports of anomalous phenomena," *Journal of Consciousness Studies* 10, no. 6-7 (2003): 153-174.

Chapter 7: Near-Death Experiences — Consciousness Beyond the Brain?

1. Sam Parnia et al., "AWARE—AWAreness during REsuscitation—A Prospective Study," *Resuscitation* 85, no. 12 (December 2014): 1799-1805.

2. Ibid., 1799.

3. Bruce Greyson, "The Near-Death Experience Scale: Construction, Reliability, and Validity," *Journal of Nervous and Mental Disease* 171, no. 6 (1983): 369.

4. Ibid., 369-375.

5. Pim van Lommel et al., "Near-Death Experience in Survivors of Cardiac Arrest: A Prospective Study in the Netherlands," *The Lancet* 358, no. 9298 (December 2001): 2039.

6. Ibid., 2039-2045.

7. Satwant Pasricha and Ian Stevenson, "Near-Death Experiences in India: A Preliminary Report," *Journal of Nervous and Mental Disease* 174, no. 3 (1986): 165-170.

8. Gregory Shushan, *Near-Death Experience in Indigenous Religions* (Oxford: Oxford University Press, 2018), 112-145.

9. Gregory Shushan, "The Historical Presence of the Near-Death Experience," in *The Handbook of Near-Death Experiences*, ed. Janice Miner Holden, Bruce Greyson, and Debbie James (Santa Barbara: Praeger, 2009), 23-42.

10. Susan Blackmore, *Dying to Live: Near-Death Experiences* (Buffalo, NY: Prometheus Books, 1993), 5-12.

11. Ibid., 85-142.

12. Gerald Woerlee, *Mortal Minds: The Biology of Near-Death Experiences* (Buffalo, NY: Prometheus Books, 2005), 127-156.

13. Steven Laureys et al., "Death, Unconsciousness and the Brain," *Nature Reviews Neuroscience* 5 (2004): 1020-1021.

14. Chris French, "Near-Death Experiences in Cardiac Arrest Survivors," *Progress in Brain Research* 150 (2005): 351-367.

15. Kevin Nelson, "Does the Arousal System Contribute to Near Death Experience?" *Neurology* 66, no. 7 (2006): 1003-1009.

16. Parnia et al., "AWARE Study," 1802-1803.

17. Bruce Greyson, "Consistency of Near-Death Experience Accounts over Two Decades: Are Reports Embellished over Time?" *Resuscitation* 73, no. 3 (2007): 407-411.

18. van Lommel et al., "Near-Death Experience in Survivors," 2044.

19. Parnia et al., "AWARE Study," 1804.

20. Anil Seth, "Consciousness and the Predictive Brain," in *Open MIND*, ed. Thomas Metzinger and Jennifer Windt (Frankfurt: MIND Group, 2015), 35(T).

21. David Chalmers, "Facing Up to the Problem of Consciousness," *Journal of Consciousness Studies* 2, no. 3 (1995): 200-219.

22. Christof Koch, "The Quest for Consciousness: A Neurobiological Approach" (Englewood, CO: Roberts & Company, 2004), 315-320.

23. Thomas Nagel, *Mind and Cosmos: Why the Materialist Neo-Darwinian Conception of Nature Is Almost Certainly False* (Oxford: Oxford University Press, 2012), 53-65.

24. Henry Stapp, "Quantum Theory and the Role of Mind in Nature," *Foundations of Physics* 31, no. 10 (2001): 1465-1499.

25. Sean Carroll, "Quantum Mechanics and the Many-Worlds Interpretation," in *The Big Picture* (New York: Dutton, 2016), 340-355.

26. Roger Penrose and Stuart Hameroff, "Consciousness in the Universe: Neuroscience, Quantum Space-Time Geometry and Orch OR Theory," *Journal of Cosmology* 14 (2011): 1-17.

27. Karl Friston, "The Free-Energy Principle: A Unified Brain Theory?" *Nature Reviews Neuroscience* 11 (2010): 127-138.

28. Steven Strogatz, *Sync: How Order Emerges from Chaos in the Universe, Nature, and Daily Life* (New York: Hyperion, 2003), 233-250.

29. Robin Carhart-Harris et al., "The Entropic Brain: A Theory of Conscious States Informed by Neuroimaging

Research with Psychedelic Drugs," *Frontiers in Human Neuroscience* 8 (2014): 20.

30. Strogatz, *Sync*, 241-245.

31. Bruce Greyson, "After-Effects of Near-Death Experiences: A Survival Mechanism Hypothesis," in *The Handbook of Near-Death Experiences*, ed. Janice Miner Holden, Bruce Greyson, and Debbie James (Santa Barbara: Praeger, 2009), 41-62.

32. Thomas Kuhn, *The Structure of Scientific Revolutions*, 4th ed. (Chicago: University of Chicago Press, 2012), 111-135.

33. Sam Parnia et al., "Guidelines and Standards for the Study of Death and Recalled Experiences of Death—A Multidisciplinary Consensus Statement and Proposed Future Directions," *Annals of the New York Academy of Sciences* 1511, no. 1 (2022): 5-21.

Chapter 8: Reincarnation — Memory Without the Self

1. Jim B. Tucker, *Return to Life: Extraordinary Cases of Children Who Remember Past Lives* (New York: St. Martin's Press, 2013), 1-18.

2. Ibid., 15-18.

3. Paul Edwards, "The Case Against Reincarnation," in *Reincarnation: A Critical Examination* (Amherst, NY: Prometheus Books, 1996), 11-35.

4. Ian Stevenson, "The Explanatory Value of the Idea of Reincarnation," *Journal of Nervous and Mental Disease* 164, no. 5 (1977): 305-326.

5. Jim B. Tucker, "A Scale to Measure the Strength of Children's Claims of Previous Lives: Methodology and Initial Findings," *Journal of Scientific Exploration* 14, no. 4 (2000): 571-581.

6. Ian Stevenson, *Where Reincarnation and Biology Intersect* (Westport, CT: Praeger, 1997), 181-195.

7. Ian Stevenson, "Birthmarks and Birth Defects Corresponding to Wounds on Deceased Persons," *Journal of Scientific Exploration* 7, no. 4 (1993): 403-410.

8. Satwant Pasricha, *Claims of Reincarnation: An Empirical Study of Cases in India* (New Delhi: Harman Publishing House, 1990), 85-120.

9. Ian Stevenson, "The Case of Imad Elawar," in *Twenty Cases Suggestive of Reincarnation*, 2nd ed. (Charlottesville: University Press of Virginia, 1974), 268-292.

10. Antonia Mills, "A Replication Study: Three Cases of Children in Northern India Who Are Said to Remember a Previous Life," *Journal of Scientific Exploration* 3, no. 2 (1989): 133-184.

11. Paul Edwards, "The Case Against Reincarnation," *Philosophical Quarterly* 15, no. 61 (1965): 37-49.

12. Susan Blackmore, "Near-Death Experiences: In or Out of the Body?" *Skeptical Inquirer* 16, no. 1 (1991): 34-45.

13. Nicholas P. Spanos et al., "Secondary Identity Enactments During Hypnotic Past-Life Regression: A Sociocognitive Perspective," *Journal of Personality and Social Psychology* 61, no. 2 (1991): 308-320.

14. Christopher C. French, "Dying to Know the Truth: Visions of a Dying Brain, or False Memories?" *Lancet* 358, no. 9298 (2001): 2010-2011.

15. Richard Wiseman, "The Luck Factor: The Scientific Study of the Lucky Mind" (London: Century, 2003), 145-167.

16. Harold I. Lief, "Commentary on Dr. Ian Stevenson's 'The Evidence for Survival from Claimed Memories of Former Incarnations,'" *Journal of Nervous and Mental Disease* 165, no. 3 (1977): 171-173.

17. Tucker, "A Scale to Measure the Strength of Children's Claims," 571-581.

18. Ian Stevenson, *Reincarnation and Biology: A Contribution to the Etiology of Birthmarks and Birth Defects*(Westport, CT: Praeger, 1997).

19. Emily Williams Kelly et al., "Can Mind Affect Matter Via Active Information?" *Mind and Matter* 5, no. 1 (2007): 59-90.

20. Elizabeth F. Loftus, "The Reality of Repressed Memories," *American Psychologist* 48, no. 5 (1993): 518-537.

21. David J. Chalmers, "Facing Up to the Problem of Consciousness," *Journal of Consciousness Studies* 2, no. 3 (1995): 200-219.

22. Robert Almeder, *Death and Personal Survival: The Evidence for Life After Death* (Lanham, MD: Rowman & Littlefield, 1992), 25-53.

23. Christof Koch, *Consciousness: Confessions of a Romantic Reductionist* (Cambridge, MA: MIT Press, 2012), 119-135.

24. Henry P. Stapp, "Quantum Theory and the Role of Mind in Nature," *Foundations of Physics* 31, no. 10 (2001): 1465-1499.

25. Stuart Hameroff, "Quantum Computation in Brain Microtubules? The Penrose–Hameroff 'Orch OR' Model of Consciousness," *Philosophical Transactions of the Royal Society A* 356, no. 1743 (1998): 1869-1896.

26. Sean Carroll, "Quantum Mechanics and the Many-Worlds Interpretation," in *The Big Picture* (New York: Dutton, 2016), 340-355.

27. Max Tegmark, "Importance of Quantum Decoherence in Brain Processes," *Physical Review E* 61, no. 4 (2000): 4194-4206.

28. Douglas R. Hofstadter, *I Am a Strange Loop* (New York: Basic Books, 2007), 201-225.

29. Rupert Sheldrake, *The Presence of the Past: Morphic Resonance and the Habits of Nature* (Rochester, VT: Park Street Press, 1988), 108-135.

30. Rachel Yehuda et al., "Holocaust Exposure Induced Intergenerational Effects on FKBP5 Methylation," *Biological Psychiatry* 80, no. 5 (2016): 372-380.

Chapter 9: Dreams, Precognition, and Alternate Selves

1. John C. Barker, "Premonitions of the Aberfan Disaster," *Journal of the Society for Psychical Research* 44, no. 732 (1967): 169-181.

2. Ibid., 169.

3. Ibid., 169-181.

4. Louisa E. Rhine, *Hidden Channels of the Mind* (New York: William Sloane Associates, 1961), 85-120.

5. Dean Radin, "Electrodermal Presentiments of Future Emotions," *Journal of Scientific Exploration* 11, no. 2 (1997): 163-180.

6. C.A. Meier, *Healing Dream and Ritual: Ancient Incubation and Modern Psychotherapy* (Einsiedeln: Daimon Verlag, 1987), 45-78.

7. Barbara Tedlock, "Dreaming and Dream Research," in *Dreaming: Anthropological and Psychological Interpretations*, ed. Barbara Tedlock (Cambridge: Cambridge University Press, 1987), 1-30.

8. Ibid., 15-25.

9. Richard Wiseman, *Paranormality: Why We See What Isn't There* (London: Macmillan, 2011), 86-105.

10. Ibid., 95-97.

11. Susan Blackmore, "Psychic Experiences: Psychic Illusions," *Skeptical Inquirer* 16, no. 4 (1992): 367-376.

12. Christopher C. French, "The Psychology of Anomalous Experiences," *Cortex* 37, no. 4 (2001): 457-469.

13. Caroline A. Watt, "Parapsychology's Contribution to Psychology: A View from the Front Line," *Journal of Parapsychology* 69, no. 2 (2005): 215-231.

14. James E. Alcock, "The Belief Engine," *Skeptical Inquirer* 19, no. 3 (1995): 14-18.

15. Dean Radin, "Unconscious Perception of Future Emotions: An Experiment in Presentiment," *Journal of Scientific Exploration* 11, no. 2 (1997): 163-180.

16. Daryl J. Bem, "Feeling the Future: Experimental Evidence for Anomalous Retroactive Influences on

Cognition and Affect," *Journal of Personality and Social Psychology* 100, no. 3 (2011): 407-425.

17. Julia A. Mossbridge et al., "Predicting the Unpredictable: Critical Analysis and Practical Implications of Predictive Anticipatory Activity," *Frontiers in Human Neuroscience* 8 (2014): 146.

18. Lynn Hasher et al., "Inhibitory Control, Circadian Arousal, and Age," in *Attention and Performance XVII*, ed. Daniel Gopher and Asher Koriat (Cambridge, MA: MIT Press, 1999), 653-675.

19. Henri Bergson, *Matter and Memory*, trans. Nancy Margaret Paul and W. Scott Palmer (London: George Allen and Unwin, 1896/1911), 170-220.

20. Julian Barbour, *The End of Time: The Next Revolution in Physics* (Oxford: Oxford University Press, 1999), 255-285.

21. Christof Koch, *Consciousness: Confessions of a Romantic Reductionist* (Cambridge, MA: MIT Press, 2012), 135-155.

22. Yakir Aharonov et al., "Time Symmetry in the Quantum Process of Measurement," *Physical Review* 134, no. 6B (1964): B1410-B1416.

23. Henry P. Stapp, "Quantum Theory and the Role of Mind in Nature," *Foundations of Physics* 31, no. 10 (2001): 1465-1499.

24. Sean Carroll, "Quantum Mechanics and the Many-Worlds Interpretation," in *The Big Picture* (New York: Dutton, 2016), 340-355.

25. Max Tegmark, "Importance of Quantum Decoherence in Brain Processes," *Physical Review E* 61, no. 4 (2000): 4194-4206.

26. Andy Clark, *Surfing Uncertainty: Prediction, Action, and the Embodied Mind* (Oxford: Oxford University Press, 2016), 190-220.

27. Rupert Sheldrake, *The Sense of Being Stared At: And Other Unexplained Powers of Human Minds* (New York: Crown Publishers, 2003), 200-225.

28. Matthew Walker, *Why We Sleep: Unlocking the Power of Sleep and Dreams* (New York: Scribner, 2017), 195-230.[9] Henri Bergson, *Matter and Memory*, trans. Nancy Margaret Paul and W. Scott Palmer (London: George Allen and Unwin, 1896/1911), 170-220.

Chapter 10: Telepathy, Collective Minds, and Synchronicity

1. Susan Blackmore, "The Elusive Open Mind: Ten Years of Negative Research in Parapsychology," *Skeptical Inquirer* 11, no. 3 (1987): 244-255.

2. Edmund Gurney, Frederic W.H. Myers, and Frank Podmore, *Phantasms of the Living*, 2 vols. (London: Trübner & Co., 1886), vol. 1, 25-35.

3. Carl G. Jung, *Synchronicity: An Acausal Connecting Principle* (Princeton: Princeton University Press, 1973), 23-51.

4. Lynne McTaggart, *The Field: The Quest for the Secret Force of the Universe* (New York: HarperCollins, 2002), 180-195.

5. Georg Feuerstein, *The Yoga Tradition: Its History, Literature, Philosophy and Practice* (Prescott, AZ: Hohm Press, 1998), 345-367.

6. Richard Katz, *Boiling Energy: Community Healing Among the Kalahari Kung* (Cambridge, MA: Harvard University Press, 1982), 120-145.

7. Edmund Gurney et al., *Phantasms of the Living*, vol. 1, 45-78.

8. Joseph Banks Rhine, *Extra-Sensory Perception* (Boston: Bruce Humphries, 1934), 85-120.

9. Charles Honorton, "Psi and Internal Attention States," in *Handbook of Parapsychology*, ed. Benjamin B. Wolman (New York: Van Nostrand Reinhold, 1977), 435-472.

10. Ray Hyman, "The Ganzfeld Psi Experiment: A Critical Appraisal," *Journal of Parapsychology* 49, no. 1 (1985): 3-49.

11. James E. Alcock, "Parapsychology: Science of the Anomalous or Search for the Soul?" *Behavioral and Brain Sciences* 10, no. 4 (1987): 553-565.

12. Christopher C. French, "Fantastic Memories: The Relevance of Research into Eyewitness Testimony and False Memories for Reports of Anomalous Experiences," *Journal of Consciousness Studies* 10, no. 6-7 (2003): 153-174.

13. Jessica Utts, "An Assessment of the Evidence for Psychic Functioning," *Journal of Scientific Exploration* 10, no. 1 (1996): 3-30.

14. Richard Wiseman and Julie Milton, "Experiment One of the SAIC Remote Viewing Program: A Critical Re-

evaluation," *Journal of Parapsychology* 62, no. 4 (1998): 297-308.

15. Caroline A. Watt, "Parapsychology's Contribution to Psychology: A View from the Front Line," *Journal of Parapsychology* 69, no. 2 (2005): 215-231.

16. Dean Radin, *The Conscious Universe: The Scientific Truth of Psychic Phenomena* (New York: HarperCollins, 1997), 85-125.

17. Daryl J. Bem, "Feeling the Future: Experimental Evidence for Anomalous Retroactive Influences on Cognition and Affect," *Journal of Personality and Social Psychology* 100, no. 3 (2011): 407-425.

18. Rupert Sheldrake, *A New Science of Life: The Hypothesis of Morphic Resonance* (Rochester, VT: Park Street Press, 1981), 95-120.

19. Andy Clark, *Surfing Uncertainty: Prediction, Action, and the Embodied Mind* (Oxford: Oxford University Press, 2016), 220-245.

20. David J. Chalmers, "Facing Up to the Problem of Consciousness," *Journal of Consciousness Studies* 2, no. 3 (1995): 200-219.

21. Giulio Tononi, "Integrated Information Theory of Consciousness: An Updated Account," *Archives Italiennes de Biologie* 150, no. 2-3 (2012): 56-90.

22. Christof Koch, *Consciousness: Confessions of a Romantic Reductionist* (Cambridge, MA: MIT Press, 2012), 155-175.

23. Henry P. Stapp, "Quantum Theory and the Role of Mind in Nature," *Foundations of Physics* 31, no. 10 (2001): 1465-1499.

24. Gregory D. Scholes et al., "Using Coherence to Enhance Function in Chemical and Biophysical Systems," *Nature*543, no. 7647 (2017): 647-656.

25. Sean Carroll, "Quantum Mechanics and the Many-Worlds Interpretation," in *The Big Picture* (New York: Dutton, 2016), 340-355.

26. Max Tegmark, "Importance of Quantum Decoherence in Brain Processes," *Physical Review E* 61, no. 4 (2000): 4194-4206.

27. Craig W. Reynolds, "Flocks, Herds and Schools: A Distributed Behavioral Model," *ACM SIGGRAPH Computer Graphics* 21, no. 4 (1987): 25-34.

28. Steven H. Strogatz, *Sync: How Order Emerges from Chaos in the Universe, Nature, and Daily Life* (New York: Hyperion, 2003), 1-25.

29. Mauricio Barahona and Louis M. Pecora, "Synchronization in Small-World Systems," *Physical Review Letters* 89, no. 5 (2002): 054101.

30. Uri Hasson et al., "Brain-to-Brain Coupling: A Mechanism for Creating and Sharing a Social World," *Trends in Cognitive Sciences* 16, no. 2 (2012): 114-121.

Chapter 11: Psychedelic Encounters and Non-Human Intelligences

1. Rick Strassman, *DMT: The Spirit Molecule* (Rochester, VT: Park Street Press, 2001), 39-45.

2. Ibid., 46-52.

3. Rick Strassman et al., "Dose-Response Study of N,N-dimethyltryptamine in Humans," *Archives of General Psychiatry* 51, no. 2 (1994): 85-97.

4. Terence McKenna, *True Hallucinations* (San Francisco: HarperSanFrancisco, 1993), 89-125.

5. Luis Eduardo Luna, "The Concept of Plants as Teachers Among Four Mestizo Shamans of Iquitos, Northeastern Peru," *Journal of Ethnopharmacology* 11, no. 2 (1984): 135-156.

6. Luis Eduardo Luna and Pablo Amaringo, *Ayahuasca Visions: The Religious Iconography of a Peruvian Shaman*(Berkeley, CA: North Atlantic Books, 1991), 45-78.

7. Omer Stewart, *Peyote Religion: A History* (Norman: University of Oklahoma Press, 1987), 89-120.

8. W.Y. Evans-Wentz, *The Tibetan Book of the Dead* (Oxford: Oxford University Press, 1927), 156-189.

9. James Fernandez, *Bwiti: An Ethnography of the Religious Imagination in Africa* (Princeton: Princeton University Press, 1982), 234-267.

10. Rick Strassman, "Human Psychopharmacology of N,N-dimethyltryptamine," *Behavioural Brain Research* 73, no. 1-2 (1996): 121-124.

11. Robin Carhart-Harris et al., "Neural Correlates of the Psychedelic State as Determined by fMRI Studies with Psilocybin," *Proceedings of the National Academy of Sciences* 109, no. 6 (2012): 2138-2143.

12. Adolf Dittrich, "The Standardized Psychometric Assessment of Altered States of Consciousness (ASCs) in

Humans," *Pharmacopsychiatry* 31, Supplement 2 (1998): 80-84.

13. Robin Carhart-Harris et al., "Psilocybin with Psychological Support for Treatment-Resistant Depression: An Open-Label Feasibility Study," *The Lancet Psychiatry* 3, no. 7 (2016): 619-627.

14. Susan Greenfield, *The Private Life of the Brain* (London: Allen Lane, 2000), 145-167.

15. Michael Persinger, "Religious and Mystical Experiences as Artifacts of Temporal Lobe Function: A General Hypothesis," *Perceptual and Motor Skills* 57, no. 3 (1983): 1255-1262.

16. Robin Carhart-Harris, "The Entropic Brain: A Theory of Conscious States Informed by Neuroimaging Research with Psychedelic Drugs," *Frontiers in Human Neuroscience* 8 (2014): 20.

17. David Luke, "Discarnate Entities and Dimethyltryptamine (DMT): Psychopharmacology, Phenomenology and Ontology," *Journal of the Society for Psychical Research* 75, no. 902 (2011): 26-42.

18. Christopher Timmermann et al., "DMT Models the Near-Death Experience," *Frontiers in Psychology* 9 (2018): 1424.

19. Benny Shanon, *The Antipodes of the Mind: Charting the Phenomenology of the Ayahuasca Experience* (Oxford: Oxford University Press, 2002), 270-295.

20. Rick Strassman, *DMT and the Soul of Prophecy* (Rochester, VT: Park Street Press, 2014), 89-120.

21. Graham Hancock, *Supernatural: Meetings with the Ancient Teachers of Mankind* (London: Century, 2005), 234-267.

22. Stanislav Grof, *LSD: Doorway to the Numinous* (Rochester, VT: Park Street Press, 2009), 156-189.

23. Dennis McKenna, *The Brotherhood of the Screaming Abyss* (St. Cloud, MN: North Star Press, 2012), 189-215.

24. Andy Clark, *Surfing Uncertainty: Prediction, Action, and the Embodied Mind* (Oxford: Oxford University Press, 2016), 245-270.

25. David J. Chalmers, "Facing Up to the Problem of Consciousness," *Journal of Consciousness Studies* 2, no. 3 (1995): 200-219.

26. Giulio Tononi, "Integrated Information Theory of Consciousness: An Updated Account," *Archives Italiennes de Biologie* 150, no. 2-3 (2012): 56-90.

27. Philip Goff, *Consciousness and Fundamental Reality* (Oxford: Oxford University Press, 2017), 178-203.

28. Christof Koch, *Consciousness: Confessions of a Romantic Reductionist* (Cambridge, MA: MIT Press, 2012), 175-195.

29. Henry P. Stapp, "Quantum Theory and the Role of Mind in Nature," *Foundations of Physics* 31, no. 10 (2001): 1465-1499.

30. Stuart Hameroff and Roger Penrose, "Consciousness in the Universe: A Review of the 'Orch OR' Theory," *Physics of Life Reviews* 11, no. 1 (2014): 39-78.

31. Sean Carroll, "Quantum Mechanics and the Many-Worlds Interpretation," in *The Big Picture* (New York: Dutton, 2016), 340-355.

32. David Chalmers, "The Conscious Mind in Search of a Fundamental Theory," *Philosophy of Mind: Classical and Contemporary Readings*, ed. David Chalmers (Oxford: Oxford University Press, 2002), 247-272.

33. Carl Gustav Jung, *Memories, Dreams, Reflections* (New York: Vintage Books, 1961), 170-199.

34. William James, *The Varieties of Religious Experience* (New York: Longmans, Green & Co., 1902), 388-420.

35. Steven Johnson, *Emergence: The Connected Lives of Ants, Brains, Cities, and Software* (New York: Scribner, 2001), 45-78.

36. Dennis McKenna, "Ayahuasca and Human Destiny," *Journal of Psychoactive Drugs* 37, no. 2 (2005): 231-234.

37. Michael Mithoefer et al., "The Safety and Efficacy of ±3,4-methylenedioxymethamphetamine-Assisted Psychotherapy in Subjects with Chronic, Treatment-Resistant Posttraumatic Stress Disorder," *Journal of Psychopharmacology* 25, no. 4 (2011): 439-452.

38. Roland Griffiths et al., "Psilocybin Produces Substantial and Sustained Decreases in Depression and Anxiety in Patients with Life-Threatening Cancer," *Journal of Psychopharmacology* 30, no. 12 (2016): 1181-1197.

Chapter 12: Savant Abilities and Sudden Genius

1. Oliver Sacks, *Musicophilia: Tales of Music and the Brain* (New York: Knopf, 2007), 3-22.

2. Ibid., 8-15.

3. John Langdon Down, "Some Observations on an Ethnic Classification of Idiots," *Journal of Mental Science* 13, no. 61 (1867): 121-123.

4. Darold A. Treffert, "Savant Syndrome: An Extraordinary Condition. A Synopsis: Past, Present, Future," *Philosophical Transactions of the Royal Society B* 364, no. 1522 (2009): 1351-1357.

5. Darold A. Treffert and Daniel D. Christensen, "Inside the Mind of a Savant," *Scientific American* 292, no. 6 (2005): 108-113.

6. Berit Brogaard and Kristian Marlow, "Profiles of Savant Syndrome," in *The Wiley Handbook of Genius*, ed. Dean Keith Simonton (Hoboken, NJ: John Wiley & Sons, 2014), 263-280.

7. Richard J. Katz, "Acceptance of Disability in Traditional Cultures," *Rehabilitation Psychology* 26, no. 3 (1981): 153-155.

8. Arthur Kleinman, *Patients and Healers in the Context of Culture* (Berkeley: University of California Press, 1980), 145-167.

9. Darold A. Treffert, *Islands of Genius: The Bountiful Mind of the Autistic, Acquired, and Sudden Savant* (London: Jessica Kingsley Publishers, 2010), 25-45.

10. Patricia Howlin et al., "Savant Skills in Autism: Psychometric Approaches and Parental Reports," *Philosophical Transactions of the Royal Society B* 364, no. 1522 (2009): 1359-1367.

11. Allan W. Snyder et al., "Savant-Like Numerosity Skills Revealed in Normal People by Magnetic Pulses," *Perception* 35, no. 6 (2006): 837-845.

12. Allan W. Snyder, "Explaining and Inducing Savant Skills: Privileged Access to Lower Level, Less-Processed Information," *Philosophical Transactions of the Royal Society B* 364, no. 1522 (2009): 1399-1405.

13. Richard P. Chi and Allan W. Snyder, "Facilitate Insight by Non-Invasive Brain Stimulation," *PLoS ONE* 6, no. 2 (2011): e16655.

14. Darold A. Treffert, "The Savant Syndrome: An Extraordinary Condition," *Philosophical Transactions of the Royal Society B* 364, no. 1522 (2009): 1351-1357.

15. Alvaro Pascual-Leone et al., "The Plastic Human Brain Cortex," *Annual Review of Neuroscience* 28 (2005): 377-401.

16. Berit Brogaard, "What Do We Learn About Consciousness from Studying the Savant Brain?" *Frontiers in Psychology* 9 (2018): 2618.

17. Allan W. Snyder, "The Savant Brain—It's the Way We All Think?" *Medical Hypotheses* 73, no. 2 (2009): 259-262.

18. Teresa Schubert and Torkel Klingberg, "Functional Neuroanatomy of Response Suppression in the Stop Signal Task," *Neuroimage* 44, no. 2 (2009): 473-481.

19. Rupert Sheldrake, *A New Science of Life: The Hypothesis of Formative Causation* (Rochester, VT: Park Street Press, 1981), 95-120.

20. Dean Radin, *Real Magic: Ancient Wisdom, Modern Science, and a Guide to the Secret Power of the Universe* (New York: Harmony Books, 2018), 145-167.

21. Jakob Hohwy, *The Predictive Mind: Cognitive Science and Philosophy of Mind* (Oxford: Oxford University Press, 2013), 189-215.

22. Stanislas Dehaene, *Consciousness and the Brain: Deciphering How the Brain Codes Our Thoughts* (New York: Viking, 2014), 156-189.

23. Giulio Tononi, "Integrated Information Theory of Consciousness: An Updated Account," *Archives Italiennes de Biologie* 150, no. 2-3 (2012): 56-90.

24. Stuart Hameroff and Roger Penrose, "Consciousness in the Universe: A Review of the 'Orch OR' Theory," *Physics of Life Reviews* 11, no. 1 (2014): 39-78.

25. Henry P. Stapp, "Quantum Theory and the Role of Mind in Nature," *Foundations of Physics* 31, no. 10 (2001): 1465-1499.

26. Max Tegmark, "Importance of Quantum Decoherence in Brain Processes," *Physical Review E* 61, no. 4 (2000): 4194-4206.

27. David Silver et al., "Mastering the Game of Go with Deep Neural Networks and Tree Search," *Nature* 529, no. 7587 (2016): 484-489.

28. Steven H. Strogatz, "Exploring Complex Networks," *Nature* 410, no. 6825 (2001): 268-276.

29. Olaf Sporns, *Networks of the Brain* (Cambridge, MA: MIT Press, 2010), 145-178.

30. Allan W. Snyder et al., "Savant-Like Skills Exposed in Normal People by Suppressing the Left Fronto-Temporal Lobe," *Journal of Integrative Neuroscience* 2, no. 2 (2003): 149-158. Brains, Cities, and Software* (New York: Scribner, 2001), 45-78.

Chapter 13: Living with Mystery

1. Open Science Collaboration, "Estimating the Reproducibility of Psychological Science," *Science* 349, no. 6251 (2015): aac4716.
2. Marcus R. Munafò et al., "A Manifesto for Reproducible Science," *Nature Human Behaviour* 1, no. 1 (2017): 0021.
3. Clifford Geertz, *The Interpretation of Cultures* (New York: Basic Books, 1973), 3-30.
4. Helen E. Longino, *Science as Social Knowledge* (Princeton: Princeton University Press, 1990), 67-82.
5. Richard P. Feynman, *The Pleasure of Finding Things Out* (Cambridge: Perseus Publishing, 1999), 240.
6. Karl R. Popper, *Of Clouds and Clocks: An Approach to the Problem of Rationality and the Freedom of Man* (St. Louis: Washington University Press, 1966).
7. Thomas S. Kuhn, *The Structure of Scientific Revolutions*, 4th ed. (Chicago: University of Chicago Press, 2012).
8. Paul K. Feyerabend, *Against Method*, 4th ed. (London: Verso, 2010).

Chapter 14: Venn Thinking — A New Model for Knowing

1. Helen E. Longino, *Science as Social Knowledge: Values and Objectivity in Scientific Inquiry* (Princeton: Princeton University Press, 1990), 67-82.
2. James D. Watson, *The Double Helix: A Personal Account of the Discovery of the Structure of DNA* (New York: Atheneum, 1968); Brenda Maddox, *Rosalind Franklin:*

The Dark Lady of DNA (New York: HarperCollins, 2002).

3. Henry Frankel, *The Continental Drift Controversy* (Cambridge: Cambridge University Press, 2012); Naomi Oreskes, *The Rejection of Continental Drift: Theory and Method in American Earth Science* (Oxford: Oxford University Press, 1999).

4. Norman Doidge, *The Brain That Changes Itself* (New York: Viking, 2007); Michael Merzenich, *Soft-Wired: How the New Science of Brain Plasticity Can Change Your Life* (San Francisco: Parnassus Publishing, 2013).

5. Mary Douglas, *How Institutions Think* (Syracuse: Syracuse University Press, 1986), 45-62.

6. John W. Creswell and Vicki L. Plano Clark, *Designing and Conducting Mixed Methods Research*, 3rd ed. (Los Angeles: SAGE Publications, 2017); Donald T. Campbell and Donald W. Fiske, "Convergent and Discriminant Validation by the Multitrait-Multimethod Matrix," *Psychological Bulletin* 56, no. 2 (1959): 81-105.

7. Stuart Hameroff and Roger Penrose, "Consciousness in the Universe: A Review of the 'Orch OR' Theory," *Physics of Life Reviews* 11, no. 1 (2014): 39-78; Max Tegmark, "Importance of Quantum Decoherence in Brain Processes," *Physical Review E* 61, no. 4 (2000): 4194-4206.

8. Thomas Nagel, *Mind and Cosmos: Why the Materialist Neo-Darwinian Conception of Nature Is Almost Certainly False* (Oxford: Oxford University Press, 2012), 15-30.

9. Mary Douglas, *How Institutions Think*, 91-109.

Afterword: Wonder as Method

1. Nathaniel C. Comfort, *The Tangled Field: Barbara McClintock's Search for the Patterns of Genetic Control*(Cambridge: Harvard University Press, 2001), 178-205.
2. Charles Darwin, *The Voyage of the Beagle* (London: John Murray, 1845), 421.
3. Jocelyn Bell Burnell, "Little Green Men, White Dwarfs, or Pulsars?" *Cosmic Search* 1, no. 1 (1979): 13-16.
4. Albert Einstein, "The World as I See It," *Forum and Century* 84, no. 4 (1930): 193-194.
5. Dacher Keltner, *Awe: The New Science of Everyday Wonder and How It Can Transform Your Life* (New York: Penguin Press, 2023), 89-112.
6. Michiel van Elk et al., "The Neural Correlates of the Awe Experience: Reduced Default Mode Network Activity during Feelings of Awe," *Human Brain Mapping* 40, no. 12 (2019): 3561-3574.
7. Paul K. Piff et al., "Awe, the Small Self, and Prosocial Behavior," *Journal of Personality and Social Psychology*108, no. 6 (2015): 883-899.
8. Paul Ricoeur, *The Symbolism of Evil*, trans. Emerson Buchanan (Boston: Beacon Press, 1967), 347-357.
9. Lee Smolin, *The Trouble with Physics: The Rise of String Theory, the Fall of a Science, and What Comes Next*(Boston: Houghton Mifflin, 2006), 285-312.
10. Rainer Maria Rilke, *Letters to a Young Poet*, trans. M.D. Herter Norton (New York: W.W. Norton, 1954), 35.

Appendix A: Glossary of Terms & Concepts

Active Inference
A theory in neuroscience proposing that the brain continuously generates predictions about sensory input and acts to minimize the difference (prediction error) between those predictions and actual sensory data. For example, when you reach for a coffee cup, your brain predicts the weight and adjusts muscle tension accordingly; unexpected lightness generates prediction error that updates future predictions. *See also: Predictive Processing*

Afterlife Traditions
Cultural or religious beliefs about existence beyond physical death, ranging from reincarnation (Hinduism, Buddhism) to resurrection (Christianity, Islam) to ancestral spirits (many African traditions). Cross-cultural surveys like those conducted by the Human Relations Area Files reveal both universal themes and significant cultural variations in afterlife concepts.

Anomalous Cognition
A broad term for cognitive phenomena that seem to occur without conventional sensory input, such as telepathy, precognition, or clairvoyance. Studied primarily through laboratory experiments using statistical analysis to detect above-chance performance in information acquisition tasks. *See also: Precognition, Telepathy*

Anomaly Proponents
Researchers who argue that evidence points toward
phenomena challenging current scientific paradigms while
maintaining scientific rigor. Unlike supernatural advocates,
they don't necessarily embrace non-physical explanations but
contend that materialist models may be insufficient.
Distinguished from both physicalist skeptics and
methodological critics. *See also: Physicalist Skeptics,
Methodological Critics*

Anthropology
The study of human societies, cultures, and their development,
including beliefs, rituals, and interpretations of extraordinary
experiences. In consciousness research, anthropological
methods help distinguish universal patterns from cultural
interpretations in phenomena like mystical experiences and
near-death reports.

Artificial Intelligence (AI)
The development of machines and systems capable of
performing tasks that normally require human intelligence,
such as learning, reasoning, and perception. Modern AI
research using neural networks provides analogies for
understanding human consciousness, distributed cognition,
and emergent intelligence. *See also: Collective Intelligence,
Extended Mind Hypothesis*

Bereavement Visions
Subjective experiences of contact with deceased individuals,
reported by approximately 60% of bereaved individuals across
cultures. Studied by grief psychologists as potentially adaptive

responses to loss, while some researchers investigate possible connections to near-death experiences and afterlife beliefs.

Blindsight
A neurological condition where individuals with damage to the primary visual cortex can respond to visual stimuli without consciously perceiving them. Demonstrates the complexity of consciousness and provides evidence for multiple processing streams in perception. Important for understanding the relationship between awareness and information processing.

Chaos Theory
A branch of mathematics and science dealing with systems that are highly sensitive to initial conditions, leading to unpredictable but patterned behavior. Relevant to consciousness studies because brain dynamics exhibit chaotic properties, and small changes in neural states might produce large changes in experience. *See also: Emergence, Self-Organization*

Collective Intelligence
The shared or group intelligence that emerges from the collaboration and competition of many individuals, often enhanced by technology or networks. Examples include swarm behavior in animals, collaborative problem-solving in human groups, and distributed processing in artificial systems. *See also: Extended Mind Hypothesis, Telepathy*

Confirmation Bias
The tendency to search for, interpret, and recall information in ways that confirm pre-existing beliefs or hypotheses. A critical

methodological concern in anomalous phenomena research, where investigators must guard against selectively attending to supportive evidence while ignoring contradictory data.

Consciousness
The state of being aware of and able to think about one's own existence, sensations, and thoughts. Its nature and origin remain actively debated, with theories ranging from emergent properties of complex neural networks to fundamental features of the universe. *See also: Hard Problem of Consciousness, Panpsychism*

Cryptomnesia
The unconscious plagiarism or false recollection of information previously encountered but forgotten, leading to the mistaken belief that the information is novel or comes from an unusual source. Particularly relevant to reincarnation research, where apparent "past-life memories" might derive from forgotten exposure to historical information.

Decoherence
In quantum mechanics, the process by which quantum superpositions collapse into definite states through interaction with the environment. Critical for understanding why quantum effects typically don't persist at macroscopic scales, which challenges theories proposing quantum mechanisms in consciousness.

Effect Size
A statistical measure of the magnitude of a phenomenon, indicating practical significance beyond mere statistical

significance. In anomalous phenomena research, effect sizes are typically small but consistent across multiple studies, raising questions about both methodological artifacts and genuine weak effects.

Emergence
A process where larger, complex patterns arise from interactions among smaller or simpler components, often with properties not predictable from the parts alone. Examples include consciousness from neural networks, flocking behavior from individual bird movements, and market dynamics from individual transactions. *See also: Self-Organization, Collective Intelligence*

Epigenetics
The study of changes in gene expression that do not involve alterations to the underlying DNA sequence, often influenced by environment and experience. Potentially relevant to reincarnation research through inherited physiological responses that might be misinterpreted as past-life memories. *See also: Cryptomnesia*

Epistemic Humility
The intellectual virtue of recognizing the limitations of one's knowledge and remaining open to revision based on new evidence. Essential for productive investigation of anomalous phenomena, balancing healthy skepticism with genuine openness to surprising discoveries.

Extended Mind Hypothesis
The theory that the mind is not confined to the brain but

extends into the environment via tools, technologies, and other people. Proposed by philosophers Andy Clark and David Chalmers, this perspective suggests that smartphones, notebooks, and social networks literally become part of our cognitive apparatus. *See also: Collective Intelligence*

Free Will
The philosophical concept that individuals can make choices unconstrained by certain factors; often debated in light of neuroscience findings about unconscious decision-making and predictive processing models that suggest conscious will may be largely illusory. *See also: Predictive Processing*

Hard Problem of Consciousness
A term coined by philosopher David Chalmers to describe the difficulty of explaining how and why subjective experiences arise from physical processes. Distinguishes between "easy problems" (explaining cognitive functions) and the hard problem (explaining subjective experience itself). Central to debates about consciousness and anomalous phenomena.

IIT (Integrated Information Theory)
A theory of consciousness proposing that conscious experience corresponds to the capacity of a system to integrate information. Developed by Giulio Tononi, IIT attempts to quantify consciousness and suggests that any system with sufficient integrated information would be conscious. *See also: Panpsychism*

Many-Worlds Interpretation
A quantum mechanics hypothesis suggesting that all possible

outcomes of a quantum event actually occur in branching, parallel universes. While highly speculative, sometimes invoked to explain apparent retrocausality or precognitive phenomena through quantum mechanical processes. *See also: Observer Effect, Superposition*

Meta-Analysis
A statistical technique that combines results from multiple independent studies to identify overall patterns and effect sizes. Crucial for evaluating anomalous phenomena research, where individual studies often show small effects but meta-analyses may reveal consistent patterns across multiple investigations.

Methodological Critics
Researchers who accept that unusual phenomena occur but question the methodology and interpretation of anomalous phenomena research. They remain within conventional scientific frameworks while raising substantive critiques about research design, alternative explanations, and evidentiary standards. *See also: Physicalist Skeptics, Anomaly Proponents*

Mystical Union
A subjective experience of merging with a perceived ultimate reality, deity, or the cosmos, often described in spiritual and religious contexts. Cross-cultural studies reveal consistent phenomenological features despite varying interpretations, suggesting possible universal neurological mechanisms. *See also: Afterlife Traditions*

Near-Death Experience (NDE)
A profound psychological event that may occur during life-

threatening situations, often involving perceptions of leaving the body, traveling through a tunnel, or encountering beings of light. Reported consistently across cultures with both universal features and culture-specific imagery. *See also: Veridical Perception, Afterlife Traditions*

Neuroplasticity
The brain's ability to reorganize itself by forming new neural connections throughout life, enabling learning, adaptation, and recovery. Relevant to understanding sudden genius and acquired savant abilities, suggesting that extraordinary cognitive capacities might be accessible through specific neural changes.

Observer Effect
In physics, the principle that the act of observation can alter the phenomenon being observed; sometimes extended metaphorically to consciousness studies. In quantum mechanics, measurement appears to collapse superposition states, raising questions about the role of consciousness in physical processes. *See also: Superposition, Decoherence*

Panpsychism
The philosophical position that consciousness is a fundamental and ubiquitous feature of all matter, similar to mass or charge. Proponents argue this solves the hard problem of consciousness by eliminating the need to explain how consciousness emerges from non-conscious matter. *See also: Hard Problem of Consciousness, IIT*

Physicalist Skeptics
Researchers who deny that anomalous phenomena represent anything beyond known brain and psychological processes. They provide comprehensive naturalistic explanations and often dismiss anomalous reports as methodological artifacts, maintaining strict materialist interpretations of all claimed anomalies. *See also: Methodological Critics, Anomaly Proponents*

Placebo Effect
A measurable, observable improvement in health not attributable to an actual treatment, but rather to the patient's belief in the treatment's efficacy. Demonstrates the power of expectation and belief to influence physical processes, relevant to understanding mind-body interactions and the role of consciousness in healing.

Precognition
The claimed ability to perceive or predict events before they occur without using known sensory channels. Laboratory studies often use random number generators or card-guessing protocols to test for above-chance performance in predicting future events. *See also: Anomalous Cognition, Meta-Analysis*

Predictive Processing
A model of brain function in which perception is the brain's best guess about the causes of sensory input, constantly updated by error correction. The brain generates predictions about incoming sensory data and adjusts these predictions based on prediction errors. *See also: Active Inference, Free Will*

Publication Bias
The tendency for academic journals to preferentially publish studies with positive or significant results while rejecting studies with negative or null results. A serious concern in anomalous phenomena research, where the "file drawer effect" might create false impressions of consistent positive findings.

Quantum Entanglement
A phenomenon in which two or more particles become linked so that their states remain correlated, even across large distances. Einstein called this "spooky action at a distance." Sometimes invoked to explain apparent telepathic phenomena, though decoherence typically prevents quantum effects at macroscopic scales. *See also: Decoherence, Observer Effect*

Replication Crisis
The discovery that many published research findings across multiple scientific disciplines fail to replicate when independent researchers attempt to reproduce the original studies. Particularly relevant to anomalous phenomena research, highlighting the importance of independent confirmation and robust methodology.

Self-Organization
The process by which order and complex patterns arise spontaneously in systems without external control. Examples include crystal formation, biological development, and the emergence of consciousness from neural interactions. *See also: Emergence, Chaos Theory*

Sleep Paralysis

A temporary inability to move or speak while falling asleep or waking up, often accompanied by vivid, frightening hallucinations. Frequently cited as a naturalistic explanation for reports of supernatural encounters, alien abductions, and some anomalous experiences.

Superposition

A fundamental principle of quantum mechanics where particles can exist in multiple states simultaneously until measured or observed. The collapse of superposition states during measurement is central to quantum theories of consciousness and debates about the observer effect. *See also: Observer Effect, Decoherence*

Synchronicity

A term coined by Carl Jung to describe meaningful coincidences that are not causally related but seem connected in significance. Jung proposed this as an "acausal connecting principle" operating alongside causality, though most scientists explain apparent synchronicities through probability theory and selective attention.

Telepathy

The claimed ability to transmit thoughts, emotions, or information directly from one mind to another without using known sensory channels. Laboratory studies often test for above-chance communication between isolated individuals, though results remain controversial. *See also: Anomalous Cognition, Quantum Entanglement*

Temporal Lobe Epilepsy
A neurological condition affecting the temporal lobes of the brain, often producing complex experiential phenomena including religious visions, out-of-body experiences, and altered states of consciousness. Sometimes proposed as an explanation for mystical experiences and anomalous phenomena.

Veridical Perception
In NDE research, reports of perceiving accurate information during periods when the brain is assumed to be incapable of normal sensory processing. These cases are considered particularly important because they potentially provide objective verification of subjective experiences. *See also: Near-Death Experience, Observer Effect*

Venn Thinking
The methodological approach developed in this book involving the systematic overlap of different explanatory frameworks to reveal patterns and insights invisible to any single perspective. Distinguished from mere eclecticism by its emphasis on rigorous methodology and productive synthesis of multiple valid approaches. *See also: Epistemic Humility, Meta-Analysis*

Appendix B: Notable Scholars

Alain Aspect

(1947–) French physicist whose experiments definitively demonstrated quantum entanglement, earning him the 2022 Nobel Prize in Physics. Professor at Institut d'Optique Graduate School and École Polytechnique. His work bridges quantum mechanics and information theory, relevant to consciousness theories invoking quantum processes. *Major work: "Bell's Inequality Test: More Ideal Than Ever" (Nature, 1999)*

Andy Clark

(1957–) Scottish philosopher and cognitive scientist who developed the Extended Mind Hypothesis with David Chalmers. Professor at University of Sussex. His interdisciplinary work bridges philosophy, cognitive science, and AI, exploring how technology extends human cognition. *Major work: "Being There: Putting Brain, Body, and World Together Again" (1997)*

Anil Seth

(1972–) British neuroscientist specializing in consciousness and perception. Co-director of the Sackler Centre for Consciousness Science, University of Sussex. His work bridges neuroscience and philosophy, developing predictive processing theories of consciousness and self-awareness. *Major work: "Consciousness and the Brain: Deciphering How the Brain Codes Our Thoughts" (2017)*

Antonio Damasio

(1944–) Portuguese-American neuroscientist exploring the neural basis of emotion and consciousness. Professor of Psychology, Philosophy, and Neuroscience at the University of Southern California. His interdisciplinary approach connects neuroscience with philosophy and psychology. *Major work: "Descartes' Error: Emotion, Reason, and the Human Brain" (1994)*

Benjamin Libet

(1916–2007) American neuroscientist whose experiments on readiness potentials challenged conventional ideas of free will. Professor Emeritus, University of California, San Francisco. His work bridges neuroscience and philosophy of mind. *Major work: "Mind Time: The Temporal Factor in Consciousness" (2004)*

Bruce Greyson

(1946–) American psychiatrist known for rigorous near-death experience research. Professor Emeritus of Psychiatry and Neurobehavioral Sciences, University of Virginia. His work bridges medical science and consciousness studies using quantitative scales and longitudinal studies. *Major work: "After: A Doctor Explores What Near-Death Experiences Reveal About Life and Beyond" (2021)*

Carlo Rovelli

(1956–) Italian theoretical physicist advocating relational quantum mechanics. Professor at Aix-Marseille University and Perimeter Institute for Theoretical Physics. His work bridges physics and philosophy, proposing that reality exists only in relationships between objects. *Major work: "The Order of Time"* *(2018)*

Christof Koch
(1956–) German-American neuroscientist studying neural correlates of consciousness. Chief Scientist of the MindScope Program at the Allen Institute for Brain Science. His work bridges neuroscience, philosophy, and information theory through Integrated Information Theory. *Major work: "Consciousness: Confessions of a Romantic Reductionist" (2012)*

Darold Treffert
(1933–2020) American psychiatrist recognized for comprehensive research on savant syndrome. Practiced and consulted in Wisconsin; affiliated with the Treffert Center. His work bridged clinical psychiatry with cognitive neuroscience, documenting extraordinary abilities in developmental disabilities. *Major work: "Islands of Genius: The Bountiful Mind of the Autistic, Acquired, and Sudden Savant" (2010)*

David Chalmers
(1966–) Australian philosopher who coined the "hard problem of consciousness." Professor of Philosophy and Neural Science, New York University. His work bridges philosophy of mind with neuroscience and AI research. *Major work: "The Conscious Mind: In Search of a Fundamental Theory" (1996)*

Dean Radin
(1952–) American psychologist and parapsychologist. Chief Scientist at the Institute of Noetic Sciences (IONS). His work applies rigorous statistical methods to anomalous phenomena

research, bridging psychology and consciousness studies. *Major work: "Real Magic: Ancient Wisdom, Modern Science, and a Guide to the Secret Power of the Universe" (2018)*

Dirk Helbing

(1965–) Swiss computational social scientist. Professor of Computational Social Science, ETH Zurich. His work bridges physics, sociology, and complex systems theory, studying emergent collective behaviors and social phenomena. *Major work: "The Automation of Society Is Next: How to Survive the Digital Revolution" (2015)*

Donald Hoffman

(1955–) American cognitive psychologist. Professor Emeritus of Cognitive Sciences, University of California, Irvine. His work bridges psychology, neuroscience, and philosophy, arguing that perception is not a window onto reality but a species-specific interface. *Major work: "The Case Against Reality: Why Evolution Hid the Truth from Our Eyes" (2019)*

Eben Alexander

(1953–) American neurosurgeon and author who reported a near-death experience during bacterial meningitis. Formerly associated with Harvard Medical School. His case bridges medical science with consciousness studies and has sparked debate about NDEs in the medical community. *Major work: "Proof of Heaven: A Neurosurgeon's Journey into the Afterlife" (2012)*

Evan Thompson

(1962–) Canadian philosopher. Professor of Philosophy,

University of British Columbia. His work bridges Western philosophy with Buddhist contemplative traditions and cognitive science, developing neurophenomenology approaches to consciousness. *Major work: "Waking, Dreaming, Being: Self and Consciousness in Neuroscience, Meditation, and Philosophy" (2015)*

Giulio Tononi
(1960–) Italian-American neuroscientist, creator of Integrated Information Theory. Professor of Psychiatry, University of Wisconsin–Madison. His work bridges neuroscience, physics, and philosophy, attempting to mathematically quantify consciousness. *Major work: "Phi: A Voyage from the Brain to the Soul" (2012)*

Ian Stevenson
(1918–2007) Canadian-born psychiatrist and reincarnation researcher. Former Chair of the Department of Psychiatry, University of Virginia. His methodical approach bridged psychiatry with anthropology and cross-cultural studies, establishing rigorous protocols for investigating reincarnation claims. *Major work: "Twenty Cases Suggestive of Reincarnation" (1974)*

James Alcock
(1942–) Canadian psychologist and skeptic. Professor Emeritus of Psychology, York University. His work bridges psychology with critical thinking education, applying rigorous psychological principles to evaluate extraordinary claims. *Major work: "Parapsychology: Science or Magic?" (1981)*

Jim Tucker

(1960–) American psychiatrist continuing Ian Stevenson's reincarnation research. Professor of Psychiatry and Neurobehavioral Sciences, University of Virginia. His work bridges child psychiatry with consciousness studies, using modern investigative techniques to study childhood memories of previous lives. *Major work: "Return to Life: Extraordinary Cases of Children Who Remember Past Lives" (2013)*

John Wheeler

(1911–2008) American theoretical physicist who coined the terms "black hole" and "wormhole." Professor at Princeton University. His work bridged quantum mechanics with cosmology and consciousness, proposing participatory universe theories. *Major work: "Geons, Black Holes, and Quantum Foam: A Life in Physics" (1998)*

Joscha Bach

(1973–) German cognitive scientist and AI researcher. Currently affiliated with the Harvard Program for Evolutionary Dynamics. His work bridges AI, cognitive science, and philosophy, developing computational theories of mind and consciousness. *Major work: "Principles of Synthetic Intelligence" (2009)*

Julie Mossbridge

(1967–) American cognitive neuroscientist. Research affiliate at the Institute of Noetic Sciences and Northwestern University. Her work bridges neuroscience with consciousness research, investigating precognitive abilities and time perception. *Major work: "The Premonition Code: The Science of Precognition" (2018)*

Karl Friston
(1959–) British neuroscientist. Professor of Imaging Neuroscience, University College London; developer of the free energy principle. His work bridges neuroscience, physics, and AI, developing mathematical frameworks for understanding brain function and consciousness. *Major work: "The Free-Energy Principle: A Unified Brain Theory?" (Nature Reviews Neuroscience, 2010)*

Laura Cabrera
(1974–) Bioethicist focusing on neuroethics and emerging technologies. Associate Professor, Michigan State University. Her work bridges ethics, neuroscience, and technology studies, examining implications of consciousness research and enhancement technologies. *Major work: "Rethinking Human Enhancement: Social Enhancement and Emergent Technologies" (2015)*

Martha McClintock
(1947–) American psychologist. Professor Emerita, University of Chicago. Her work bridges psychology and biology, studying social and hormonal influences on behavior and the potential for unconscious communication between individuals. *Major work: "Menstrual Synchrony and Suppression" (Nature, 1971)*

Menno Schilthuizen
(1965–) Dutch evolutionary biologist. Senior researcher at Naturalis Biodiversity Center, Leiden, and Professor at Leiden

University. His work bridges evolutionary biology with urban ecology, studying rapid adaptation and emergent behaviors in changing environments. *Major work: "Darwin Comes to Town: How the Urban Jungle Drives Evolution" (2018)*

Michael Graziano

(1967–) American neuroscientist. Professor of Psychology and Neuroscience, Princeton University. His work bridges neuroscience with consciousness studies, developing attention schema theory as an explanation for subjective awareness. *Major work: "Consciousness and the Social Brain" (2013)*

Michael Levin

(1969–) American developmental biologist. Professor and Director, Allen Discovery Center at Tufts University. His work bridges biology with information theory and cognitive science, studying bioelectric signaling and collective intelligence in biological systems. *Major work: "The Computational Boundary of a 'Self': Developmental Bioelectricity Drives Multicellularity and Scale-Free Cognition" (2019)*

Philip Goff

(1978–) British philosopher. Professor of Philosophy, Durham University. His work bridges philosophy of mind with physics, developing panpsychist theories of consciousness and defending the combination problem. *Major work: "Consciousness and Fundamental Reality" (2017)*

Pim van Lommel

(1943–) Dutch cardiologist and NDE researcher. Retired from clinical practice; continues scholarly work. His prospective study of cardiac arrest patients bridges cardiology with consciousness research, providing rigorous medical documentation of NDEs. *Major work: "Consciousness Beyond Life: The Science of the Near-Death Experience" (2010)*

Ray Hyman
(1930–2024) American psychologist and noted skeptic. Professor Emeritus, University of Oregon. His work bridged psychology with critical thinking, developing methodological critiques of parapsychology research and promoting scientific skepticism. *Major work: "The Elusive Quarry: A Scientific Appraisal of Psychical Research" (1989)*

Richard Wiseman
(1966–) British psychologist and skeptic. Professor of Psychology, University of Hertfordshire. His work bridges psychology with magic and deception research, applying psychological principles to evaluate paranormal claims. *Major work: "Paranormality: Why We See What Isn't There" (2011)*

Rick Strassman
(1952–) American psychiatrist. Clinical Associate Professor of Psychiatry, University of New Mexico School of Medicine. His pioneering research bridges psychiatry with consciousness studies through clinical psychedelic research, particularly with DMT. *Major work: "DMT: The Spirit Molecule" (2001)*

Robin Carhart-Harris
(1980–) British neuroscientist. Ralph Metzner Distinguished Professor of Neurology and Psychiatry, University of California, San Francisco. His work bridges neuroscience with

consciousness research, using psychedelics to study brain function and develop therapeutic applications. *Major work: "How Do Psychedelics Work?" (Current Opinion in Psychiatry, 2019)*

Roger Penrose
(1931–) British mathematical physicist and Nobel laureate. Emeritus Professor of Mathematics, University of Oxford. His work bridges mathematics, physics, and consciousness studies, proposing quantum theories of mind through orchestrated objective reduction. *Major work: "The Emperor's New Mind: Concerning Computers, Minds, and the Laws of Physics" (1989)*

Roland Griffiths
(1946–2023) American psychopharmacologist. Professor in the Departments of Psychiatry and Neurosciences, Johns Hopkins University School of Medicine. His pioneering work bridged pharmacology with consciousness research, conducting rigorous clinical studies of psychedelic experiences. *Major work: "Psilocybin Can Occasion Mystical-Type Experiences" (Psychopharmacology, 2006)*

Rupert Sheldrake
(1942–) British biologist. Former Director of Studies in Cell Biology, Clare College, Cambridge University; now independent researcher and author. His controversial work bridges biology with consciousness research, proposing morphic resonance as a mechanism for collective memory. *Major work: "A New Science of Life: The Hypothesis of Morphic Resonance" (1981)*

Sabine Hossenfelder

(1976–) German theoretical physicist. Research fellow at the Frankfurt Institute for Advanced Studies. Her work bridges physics with philosophy of science, critically examining foundational assumptions in physics and consciousness theories. *Major work: "Lost in Math: How Beauty Leads Physics Astray" (2018)*

Sam Harris

(1967–) American neuroscientist and author. Holds a PhD in cognitive neuroscience from UCLA; founder of the Waking Up meditation app. His work bridges neuroscience with contemplative traditions and philosophy, studying meditation and consciousness. *Major work: "Waking Up: A Guide to Spirituality Without Religion" (2014)*

Sam Parnia

(1972–) British intensivist and resuscitation researcher. Director of Critical Care and Resuscitation Research, NYU Langone Health. His work bridges emergency medicine with consciousness research, conducting rigorous studies of consciousness during cardiac arrest. *Major work: "What Happens When We Die: A Groundbreaking Study into the Nature of Life and Death" (2006)*

Sean Carroll

(1966–) American theoretical physicist. Professor at Johns Hopkins University. His work bridges cosmology with philosophy, developing many-worlds interpretations of quantum mechanics and examining implications for consciousness and free will. *Major work: "The Big Picture: On the Origins of Life, Meaning, and the Universe Itself" (2016)*

Stanislav Grof

(1931–) Czech-American psychiatrist and pioneer in transpersonal psychology. Professor at the California Institute of Integral Studies. His work bridges psychiatry with consciousness research through psychedelic therapy and holotropic breathwork. *Major work: "Realms of the Human Unconscious: Observations from LSD Research" (1975)*

Steven Laureys

(1968–) Belgian neurologist. Professor at the University of Liège; Director of the Coma Science Group. His work bridges neurology with consciousness research, studying awareness in disorders of consciousness and developing clinical assessment tools. *Major work: "The Neurology of Consciousness" (2009)*

Steven Strogatz

(1959–) American mathematician. Professor of Applied Mathematics, Cornell University. His work bridges mathematics with complex systems theory, studying synchronization and emergent behaviors in biological and social networks. *Major work: "Sync: How Order Emerges from Chaos in the Universe, Nature, and Daily Life" (2003)*

Susan Blackmore

(1951–) British psychologist and author. Visiting Professor at the University of Plymouth. Her work bridges psychology with consciousness studies and skeptical inquiry, transitioning from parapsychology researcher to prominent skeptic. *Major work: "Consciousness: An Introduction" (2003)*

Susan Greenfield

(1950–) British neuroscientist and author. Senior Research Fellow, Lincoln College, University of Oxford. Her work bridges neuroscience with technology studies, examining how digital technology affects brain development and consciousness. *Major work: "Mind Change: How Digital Technologies Are Leaving Their Mark on Our Brains" (2015)*

Tenzin Gyatso (14th Dalai Lama)

(1935–) Tibetan Buddhist leader and Nobel Peace Prize laureate. His extensive collaborations with Western scientists bridge contemplative traditions with neuroscience, particularly through the Mind & Life Institute. His work has influenced scientific study of meditation and consciousness. *Major work: "The Universe in a Single Atom: The Convergence of Science and Spirituality" (2005)*

Thomas Metzinger

(1958–) German philosopher. Professor of Theoretical Philosophy, Johannes Gutenberg University Mainz. His work bridges philosophy with neuroscience and AI, developing sophisticated theories of selfhood and conscious experience. *Major work: "Being No One: The Self-Model Theory of Subjectivity" (2003)*

William James

(1842–1910) American philosopher and psychologist. Professor at Harvard University; pioneer in the study of religious experience and founder of American psychology. His work bridged psychology with philosophy and religious studies, establishing scientific approaches to consciousness. *Major work: "The Varieties of Religious Experience: A Study in Human Nature" (1902)*

Yann LeCun
(1960–) French computer scientist. Chief AI Scientist at Meta and Silver Professor at New York University. His pioneering work in deep learning bridges computer science with cognitive science, influencing theories of how artificial and biological intelligence might relate. *Major work: "Deep Learning" (2016)*

Yoshua Bengio
(1964–) Canadian computer scientist. Professor at the Université de Montréal and Scientific Director of Mila – Quebec AI Institute. His foundational work in AI bridges computer science with cognitive science and consciousness research. *Major work: "Learning Deep Architectures for AI" (2009)*

Appendix C: Further Reading Matrix

Introduction & Prologue — Methodology, Cultural Framing, History of the "Unexplained"

- Kuhn, Thomas S. *The Structure of Scientific Revolutions.* Chicago: University of Chicago Press, 2012. *Classic analysis of how scientific paradigms shift and anomalies drive theoretical revolution.*
- Sagan, Carl. *The Demon-Haunted World: Science as a Candle in the Dark.* New York: Ballantine, 1996. *Influential defense of scientific skepticism and critical thinking about extraordinary claims.*
- Firestein, Stuart. *Ignorance: How It Drives Science.* New York: Oxford University Press, 2012. *Argues that productive ignorance and well-crafted questions drive scientific progress.*
- Longino, Helen E. *Science as Social Knowledge: Values and Objectivity in Scientific Inquiry.* Princeton: Princeton University Press, 1990. *Foundational work on how social factors shape scientific knowledge without undermining objectivity.*

Chapter 1 — The Predictive Brain

- Friston, Karl. "The Free-Energy Principle: A Unified Brain Theory?" *Nature Reviews Neuroscience* 11, no. 2 (2010): 127-38. *Mathematical framework proposing that brains minimize surprise through predictive modeling.*

- Seth, Anil. *Being You: A New Science of Consciousness.* London: Faber & Faber, 2021. *Accessible introduction to predictive processing theories of consciousness and selfhood.*
- Clark, Andy. *Surfing Uncertainty: Prediction, Action, and the Embodied Mind.* Oxford: Oxford University Press, 2016. *Comprehensive philosophical treatment of predictive processing and its implications.*
- Harris, Sam. *Free Will.* New York: Free Press, 2012. *Neuroscience-informed argument that conscious will is largely illusory.*
- Libet, Benjamin. *Mind Time: The Temporal Factor in Consciousness.* Cambridge: Harvard University Press, 2004. *Detailed presentation of experiments challenging conventional ideas about conscious will.*

Chapter 2 — Consciousness: Fundamental or Emergent?

- Chalmers, David J. *The Conscious Mind: In Search of a Fundamental Theory.* New York: Oxford University Press, 1996. *Seminal work defining the "hard problem" of consciousness and defending property dualism.*
- Tononi, Giulio. *Phi: A Voyage from the Brain to the Soul.* New York: Pantheon, 2012. *Accessible presentation of Integrated Information Theory and its implications.*
- Koch, Christof. *The Feeling of Life Itself: Why Consciousness Is Widespread but Can't Be Computed.* Cambridge: MIT Press, 2019. *Neuroscientist's argument for panpsychist theories of consciousness.*

- Damasio, Antonio. *The Feeling of What Happens: Body and Emotion in the Making of Consciousness.* New York: Harcourt Brace, 1999. *Influential theory emphasizing embodied emotion in consciousness.*
- Dennett, Daniel C. *Consciousness Explained.* Boston: Little, Brown, 1991. *Eliminative materialist approach arguing consciousness is less mysterious than commonly assumed.*
- Goff, Philip. *Consciousness and Fundamental Reality.* Oxford: Oxford University Press, 2017. *Contemporary philosophical defense of panpsychism as solution to consciousness problems.*

Chapter 3 — Quantum Reality and Emergent Complexity

- Penrose, Roger. *The Road to Reality: A Complete Guide to the Laws of the Universe.* New York: Knopf, 2004. *Comprehensive mathematical treatment of physics including quantum consciousness theories.*
- Rovelli, Carlo. *Helgoland: Making Sense of the Quantum Revolution.* London: Allen Lane, 2021. *Relational interpretation of quantum mechanics and its philosophical implications.*
- Hossenfelder, Sabine. *Existential Physics: A Scientist's Guide to Life's Biggest Questions.* New York: Viking, 2022. *Physicist's skeptical examination of quantum consciousness theories and their limitations.*
- Mitchell, Melanie. *Complexity: A Guided Tour.* Oxford: Oxford University Press, 2009. *Accessible introduction to complex systems theory and emergence.*

- Tegmark, Max. "Importance of Quantum Decoherence in Brain Processes." *Physical Review E* 61, no. 4 (2000): 4194-4206. *Technical critique arguing quantum effects cannot persist in warm, noisy brains.*

Chapter 4 — Artificial Minds as Cognitive Mirrors

- Bengio, Yoshua, Ian Goodfellow, and Aaron Courville. *Deep Learning.* Cambridge: MIT Press, 2016. *Comprehensive technical introduction to modern AI and neural networks.*
- Bach, Joscha. "Artificial Consciousness: Artificial Intelligence and the Nature of Consciousness." *Frontiers in Psychology* 13 (2022). *Computational approach to understanding consciousness through AI modeling.*
- LeCun, Yann. "A Path Towards Autonomous Machine Intelligence." *OpenReview*, 2022. *Vision for future AI development and its relationship to human intelligence.*
- Mitchell, Melanie. *Artificial Intelligence: A Guide for Thinking Humans.* New York: Farrar, Straus and Giroux, 2019. *Critical examination of AI capabilities and limitations with implications for consciousness research.*
- Russell, Stuart. *Human Compatible: Artificial Intelligence and the Problem of Control.* New York: Viking, 2019. *Analysis of AI development challenges with implications for understanding intelligence.*

Chapter 5 — Religion, Mysticism, and the Science of Subjective Realities

- Hoffman, Donald. *The Case Against Reality: How Evolution Hid the Truth from Our Eyes.* New York: W. W. Norton, 2019. *Argument that perception evolved for fitness, not truth, with implications for mystical experience.*

- James, William. *The Varieties of Religious Experience.* New York: Longmans, Green, 1902. *Classic psychological study of religious and mystical experiences across cultures.*

- Grof, Stanislav. *The Adventure of Self-Discovery.* Albany: SUNY Press, 1988. *Systematic exploration of altered states through psychedelic therapy and breathwork.*

- Thompson, Evan. *Waking, Dreaming, Being: Self and Consciousness in Neuroscience, Meditation, and Philosophy.* New York: Columbia University Press, 2015. *Integration of Western neuroscience with Buddhist contemplative traditions.*

- Newberg, Andrew, and Mark Robert Waldman. *How God Changes Your Brain.* New York: Ballantine, 2009. *Neuroscientific study of meditation and religious experience.*

- Hood Jr., Ralph W. "The Construction and Preliminary Validation of a Measure of Reported Mystical Experience." *Journal for the Scientific Study of Religion* 14, no. 1 (1975): 29-41. *Empirical research on mystical experience measurement.*

Chapter 6 — Systems and Networks: Patterns Behind the Unexplainable

- Strogatz, Steven. *Sync: How Order Emerges from Chaos in the Universe, Nature, and Daily Life.* New York: Hyperion, 2003. *Accessible exploration of synchronization phenomena across multiple domains.*
- Helbing, Dirk. *Social Self-Organization: Agent-Based Simulations and Experiments to Study Emergent Social Behavior.* Berlin: Springer, 2012. *Technical treatment of collective behavior and emergent social phenomena.*
- Levin, Michael. "Bioelectric Signaling: Reprogrammable Circuits Underlying Morphogenesis." *Cell* 184, no. 4 (2021): 911-26. *Cutting-edge research on collective intelligence in biological systems.*
- Barabási, Albert-László. *Linked: How Everything Is Connected to Everything Else.* New York: Plume, 2003. *Introduction to network science and its applications across disciplines.*
- Watts, Duncan J. *Six Degrees: The Science of a Connected Age.* New York: W. W. Norton, 2003. *Exploration of small-world networks and their implications for social phenomena.*

Chapter 7 — Near-Death Experiences

- Greyson, Bruce. *After: A Doctor Explores What Near-Death Experiences Reveal about Life and Beyond.* New York: St. Martin's Essentials, 2021. *Leading researcher's comprehensive overview of NDE research and implications.*
- Parnia, Sam. "Death and Consciousness—An Overview of the Evidence." *Annals of the New York Academy of*

Sciences 1330 (2014): 75-93. *Medical perspective on consciousness during cardiac arrest.*

- van Lommel, Pim. *Consciousness Beyond Life: The Science of the Near-Death Experience.* New York: HarperOne, 2010. *Cardiologist's prospective study and theoretical framework.*
- Blackmore, Susan. *Dying to Live: Near-Death Experiences.* Buffalo: Prometheus Books, 1993. *Comprehensive skeptical analysis of NDE phenomena and alternative explanations.*
- Mobbs, Dean, and Caroline Watt. "There Is Nothing Paranormal About Near-Death Experiences." *Trends in Cognitive Sciences* 15, no. 10 (2011): 447-49. *Neuroscientific critique emphasizing dying brain explanations.*
- Holden, Janice Miner, Bruce Greyson, and Debbie James, eds. *The Handbook of Near-Death Experiences.* Santa Barbara: Praeger, 2009. *Comprehensive academic collection covering research methods and findings.*

Chapter 8 — Reincarnation

- Stevenson, Ian. *Twenty Cases Suggestive of Reincarnation.* Charlottesville: University Press of Virginia, 1974. *Foundational work establishing methodological standards for reincarnation research.*
- Tucker, Jim B. *Life Before Life: A Scientific Investigation of Children's Memories of Previous Lives.* New York: St. Martin's Press, 2005. *Contemporary continuation of Stevenson's work with enhanced methodology.*

- Mills, Antonia. "Reincarnation Beliefs of North American Indians and Inuit." *Anthropology of Consciousness* 8, no. 1 (1997): 47-68. *Cross-cultural analysis of reincarnation beliefs and practices.*
- Humphrey, Nicholas. *Soul Dust: The Magic of Consciousness.* Princeton: Princeton University Press, 2011. *Skeptical analysis of consciousness and survival claims.*
- Edwards, Paul. *Reincarnation: A Critical Examination.* Amherst: Prometheus Books, 1996. *Comprehensive philosophical and empirical critique of reincarnation claims.*

Chapter 9 — Dreams, Precognition, and Alternate Selves

- Mossbridge, Julia, and Dean Radin. *The Premonition Code: The Science of Precognition.* London: Watkins, 2018. *Meta-analytic review of precognition research and theoretical frameworks.*
- Barrett, Deirdre. *The Committee of Sleep.* New York: Crown, 2001. *Scientific study of problem-solving and creativity in dreams.*
- Stickgold, Robert. "Sleep-Dependent Memory Consolidation." *Nature* 437 (2005): 1272-78. *Neuroscientific foundation for understanding dream function.*
- Wagenmakers, Eric-Jan, et al. "Registered Replication Report: Bem (2011)." *Perspectives on Psychological Science* 11, no. 4 (2016): 539-44. *Failed replication of controversial precognition experiments.*

- Radin, Dean I. "Unconscious Perception of Future Emotions: An Experiment in Presentiment." *Journal of Scientific Exploration* 11, no. 2 (1997): 163-80. *Original research on precognitive physiological responses.*

Chapter 10 — Telepathy, Collective Minds, and Synchronicity

- Sheldrake, Rupert. *The Sense of Being Stared At: And Other Aspects of the Extended Mind.* New York: Crown, 2003. *Controversial research on telepathy and morphic resonance.*
- Radin, Dean. *Entangled Minds: Extrasensory Experiences in a Quantum Reality.* New York: Paraview, 2006. *Meta-analytic review of telepathy research with quantum theoretical framework.*
- Jung, Carl, and Wolfgang Pauli. *The Interpretation of Nature and the Psyche.* New York: Pantheon, 1955. *Classic work on synchronicity and its relationship to quantum physics.*
- Storm, Lance, Patrizio E. Tressoldi, and Lorenzo Di Risio. "Meta-Analysis of Free-Response Studies, 1992–2008." *Psychological Bulletin* 136, no. 4 (2010): 471-85. *Statistical analysis finding small but significant telepathy effects.*
- Wiseman, Richard, and Julie Milton. "Experiment One of the SAIC Remote Viewing Program: A Critical Re-evaluation." *Journal of Parapsychology* 62, no. 4 (1998): 297-308. *Methodological critique of remote viewing research.*

- Hyman, Ray. "Meta-Analysis That Conceals More Than It Reveals: Comment on Storm et al. (2010)." *Psychological Bulletin* 136, no. 4 (2010): 486-90. *Skeptical critique of telepathy meta-analyses.*

Chapter 11 — Psychedelic Encounters and Non-Human Intelligences

- Strassman, Rick. *DMT: The Spirit Molecule.* Rochester: Park Street Press, 2001. *Clinical research on DMT experiences and entity encounters.*
- Carhart-Harris, Robin, et al. "The Entropic Brain: A Theory of Conscious States Informed by Neuroimaging Research with Psychedelic Drugs." *Frontiers in Human Neuroscience* 8 (2014): 20. *Neuroscientific framework for understanding psychedelic effects.*
- Griffiths, Roland R., et al. "Psilocybin Can Occasion Mystical-Type Experiences Having Substantial and Sustained Personal Meaning." *Psychopharmacology* 187 (2006): 268-83. *Rigorous clinical study of psilocybin's consciousness effects.*
- Shanon, Benny. *The Antipodes of the Mind: Charting the Phenomenology of the Ayahuasca Experience.* Oxford: Oxford University Press, 2002. *Systematic phenomenological analysis of ayahuasca experiences.*
- Luke, David. "Anomalous Psychedelic Experiences: At the Neurochemical Juncture Between the Implausible and the Possible." *Frontiers in Psychology* 3 (2012): 309. *Analysis of apparently paranormal effects in psychedelic states.*

Chapter 12 — Savant Abilities and Sudden Genius

- Treffert, Darold A. *Islands of Genius: The Bountiful Mind of the Autistic, Acquired, and Sudden Savant.* London: Jessica Kingsley Publishers, 2011. *Comprehensive overview of savant syndrome across different populations.*
- Snyder, Allan. "Explaining and Inducing Savant Skills: Privileged Access to Lower Level, Less-Processed Information." *Philosophical Transactions of the Royal Society B* 364 (2009): 1399-1405. *Theoretical framework for understanding savant abilities.*
- Treffert, Darold A., and Daniel D. Christensen. "Inside the Mind of a Savant." *Scientific American Mind* 17, no. 5 (2006): 56-63. *Accessible overview of savant syndrome research.*
- Young, Richard L., et al. "The Emergence of Savant Abilities Following Left Anterior Temporal Lobe Damage." *Neuropsychologia* 43, no. 12 (2005): 1722-26. *Case study of acquired savant abilities.*
- Hou, Changbing, et al. "Sudden Savant Syndrome Following Stroke: A Case Report." *Neurocase* 26, no. 1 (2020): 38-43. *Recent documentation of sudden genius acquisition.*

Chapter 13 — Living with Mystery

- McGrayne, Sharon Bertsch. *The Theory That Would Not Die.* New Haven: Yale University Press, 2011. *History of Bayesian statistics and uncertainty quantification.*
- Ioannidis, John P. A. "Why Most Published Research Findings Are False." *PLoS Medicine* 2, no. 8 (2005): e124.

Influential analysis of publication bias and false positives in research.

- Firestein, Stuart. *Ignorance: How It Drives Science.* New York: Oxford University Press, 2012. *Argument that productive ignorance drives scientific discovery.*
- Munafò, Marcus R., et al. "A Manifesto for Reproducible Science." *Nature Human Behaviour* 1, no. 1 (2017): 0021. *Analysis of replication crisis and proposed solutions.*
- Popper, Karl R. *Conjectures and Refutations.* London: Routledge, 2002. *Classic philosophy of science emphasizing fallibilism and provisional knowledge.*

Chapter 14 — Venn Thinking: A New Model for Knowing

- Snow, C. P. *The Two Cultures.* Cambridge: Cambridge University Press, 1959. *Classic analysis of the divide between scientific and humanistic knowledge.*
- Epstein, David. *Range: Why Generalists Triumph in a Specialized World.* New York: Riverhead Books, 2019. *Argument for the value of interdisciplinary thinking and diverse experience.*
- Klein, Gary. *Seeing What Others Don't: The Remarkable Ways We Gain Insights.* New York: PublicAffairs, 2013. *Cognitive research on breakthrough thinking and insight generation.*
- Klein, Julie Thompson. *Interdisciplinarity: History, Theory, and Practice.* Detroit: Wayne State University Press, 1990. *Comprehensive analysis of interdisciplinary research methods and challenges.*

- Creswell, John W., and Vicki L. Plano Clark. *Designing and Conducting Mixed Methods Research*. Los Angeles: SAGE Publications, 2017. *Methodological guide for integrating quantitative and qualitative approaches.*

Afterword — Wonder as Method

- Abram, David. *The Spell of the Sensuous: Perception and Language in a More-Than-Human World*. New York: Vintage, 1996. *Phenomenological approach to wonder and perception in nature.*
- Keltner, Dacher. *Awe: The New Science of Everyday Wonder and How It Can Transform Your Life*. New York: Penguin Press, 2023. *Psychological research on awe and its cognitive and social benefits.*
- Rovelli, Carlo. *Seven Brief Lessons on Physics*. New York: Riverhead Books, 2016. *Poetic meditation on physics and wonder at natural phenomena.*
- Carson, Rachel. *The Sense of Wonder*. New York: Harper & Row, 1965. *Classic reflection on maintaining curiosity and awe in scientific inquiry.*
- Feynman, Richard P. *The Pleasure of Finding Things Out*. Cambridge: Perseus Publishing, 1999. *Nobel laureate's reflections on curiosity and scientific discovery.*

Digital Resources and Databases

Academic Journals and Societies:

- *Journal of Consciousness Studies* — Interdisciplinary consciousness research
- *Journal of Near-Death Studies* — Peer-reviewed NDE research
- *Journal of Scientific Exploration* — Anomalous phenomena research
- *Frontiers in Psychology: Consciousness Research* — Open-access consciousness studies
- Mind & Life Institute — Science-contemplative tradition dialogue

Research Databases:

- PubMed/MEDLINE — Biomedical literature search
- PhilPapers — Philosophy research database
- arXiv — Preprint server for physics and consciousness research
- Cochrane Library — Systematic reviews and meta-analyses

Professional Organizations:

- Association for the Scientific Study of Consciousness (ASSC)
- International Association for Near-Death Studies (IANDS)
- Society for Scientific Exploration (SSE)
- Consciousness Research Network
- Parapsychological Association

Methodological and Critical Frameworks

Research Methods:

- Campbell, Donald T., and Donald W. Fiske. "Convergent and Discriminant Validation by the Multitrait-Multimethod Matrix." *Psychological Bulletin* 56, no. 2 (1959): 81-105. *Foundational work on methodological triangulation.*
- Bem, Daryl J., et al. "Must Psychologists Change the Way They Analyze Their Data?" *Journal of Personality and Social Psychology* 101, no. 4 (2011): 716-19. *Debate over statistical methods in controversial research.*

Critical Perspectives:

- Alcock, James E. *Science and Supernature: A Critical Appraisal of Parapsychology.* Buffalo: Prometheus Books, 1990. *Comprehensive skeptical analysis of paranormal research.*
- Shermer, Michael. *Why People Believe Weird Things.* New York: W. H. Freeman, 1997. *Psychological analysis of belief in extraordinary claims.*
- Hines, Terence. *Pseudoscience and the Paranormal.* Amherst: Prometheus Books, 2003. *Critical examination of paranormal claims and research methods.*

Philosophical Foundations:

- Nagel, Thomas. *Mind and Cosmos: Why the Materialist Neo-Darwinian Conception of Nature Is Almost Certainly*

False. Oxford: Oxford University Press, 2012. *Philosophical critique of reductive materialism.*

- McGinn, Colin. *The Mysterious Flame: Conscious Minds in a Material World.* New York: Basic Books, 1999. *Analysis of consciousness as potentially beyond human understanding.*

About the Author

Kevin Meyer is a retired executive and lifelong student of subjects that challenge conventional thinking. Trained as a chemical engineer, he spent most of his career in executive roles in medical device manufacturing, where he developed a deep appreciation for the intersection of rigorous methodology and human-centered problem solving.

Meyer cofounded an e-learning company focused on continuous improvement methods, which he successfully grew and eventually sold. This experience reinforced his belief in the power of making complex ideas accessible to broader audiences—a principle that guides his current writing projects.

Born in the United States but raised for seven formative years in Peru, Meyer has traveled to over sixty-five countries, always with an eye toward understanding how different cultures approach fundamental questions about meaning, truth, and human flourishing. This global perspective shapes his approach to exploring diverse intellectual traditions with both curiosity and respect.

For nearly thirty years, Meyer has practiced *bhāvanā*—disciplined cultivation of understanding through deep, intentional study of subjects ranging from marathon running and scuba diving to philosophy, spirituality, and history. Each year, he embraces a new area of inquiry that pushes him beyond his comfort zone and challenges his assumptions.

Now retired and living in Morro Bay, California, Meyer finally has the time to fully explore the wildly divergent topics that have always fascinated him. He remains active in the local biotech startup community, bringing his experience in scaling complex operations to emerging companies.

Meyer can be reached through his website at KevinMeyer.com.

Other Books by Kevin Meyer

Sacred Editors Series

Sacred Editors: How Power, Politics, and Interpretation Shaped the Christian Scripture
The remarkable human story behind the formation of the Bible, tracing how scribes, bishops, emperors, and competing Christian communities shaped the New Testament canon through centuries of copying, translating, and theological debate.

Sacred Editors: How Exile, Law, and Dialogue Evolved Jewish Sacred Texts
An exploration of how the Hebrew Bible and Talmud came to be, examining the human decisions, political pressures, and scholarly debates that shaped Jewish sacred texts across centuries of transmission and interpretation.

Sacred Editors: How Preservation and Authority Defined Islamic Sacred Texts
The fascinating story behind the compilation of the Quran and the development of Islamic textual traditions, revealing the complex historical processes through which Muslim communities preserved and codified their foundational texts.

Sacred Editors: How Preservation, Transmission, and Insight Shaped the Buddhist Canon
The fascinating story of how the Buddha's teachings transformed from remembered words into the world's most

diverse collection of sacred canons, spanning cultures from Sri Lanka to Tibet to Japan.

Sacred Editors: How Tradition, Interpretation, and Devotion Shaped Hindu Sacred Texts

The extraordinary journey of Hindu sacred texts from ancient oral recitation to global digital archives, exploring how tradition-keepers, commentators, and communities across millennia have shaped the Vedas, Upanishads, epics, and Puranas.

Interfaith and Spiritual Exploration

The Beatitudes Path: An Interfaith Exploration of Sacred Blessings

A transformative journey through Jesus' most beloved teachings that reveals how the Beatitudes articulate universal truths about compassion, justice, and authentic living that transcend religious boundaries. Drawing on scholarship from Buddhism to Islam, Indigenous wisdom to ancient Greek philosophy, Meyer demonstrates how these ancient blessings offer profound guidance for seekers of every background wrestling with life's deepest questions.

Leadership and Personal Development

The Simple Leader: Personal and Professional Leadership at the Nexus of Lean and Zen

Filled with personal stories, practices, and insights from a thirty-year leadership journey, this book reveals the surprising connections between Lean manufacturing principles and Zen wisdom. Organized into eight practical parts—from reconnecting with your inner self to growing your organization—it shows leaders in any industry how to become more organized, effective, and balanced by integrating concepts like simplicity, flow, respect, and beginner's mind into their daily practice.

Biography

Harleigh Knott: Christmas Greetings From a Remarkable Life

A celebration of Harleigh Thayer Knott (1929-2019), a remarkable woman who lived her entire life in Morro Bay, California, while developing interests ranging from opera to history, polo to Indy car racing. Compiled from sixty years of her captivating and humorous annual Christmas letters, this book preserves the memory of an extraordinary life filled with Stanford education, world travel, and an insatiable curiosity about everything from frogs to the human condition.

www.ingramcontent.com/pod-product-compliance
Lightning Source LLC
Chambersburg PA
CBHW071204090426
42736CB00014B/2708